Society of Economic Geologists Annual Conference (SEG 2022)

Minerals for our Future

Denver, Colorado, USA
27 - 30 August 2022

ISBN: 978-1-7138-6664-0

Printed from e-media with permission by:

Curran Associates, Inc.
57 Morehouse Lane
Red Hook, NY 12571

Some format issues inherent in the e-media version may also appear in this print version.

Printed with permission by Curran Associates, Inc. (2023)

For permission requests, please contact Society of Economic Geologists (SEG)
at the address below.

Society of Economic Geologists (SEG)
7811 Shaffer Parkway
Littleton, CO 80127-3732
USA

Phone: +1(720) 981-7882
Fax: +1(720) 981-7874

www.segweb.org

Additional copies of this publication are available from:

Curran Associates, Inc.
57 Morehouse Lane
Red Hook, NY 12571 USA
Phone: 845-758-0400
Fax: 845-758-2633
Email: curran@proceedings.com
Web: www.proceedings.com

TABLE OF CONTENTS

CRITICAL MINERALS FOR OUR ENERGY FUTURE: GEOLOGY AND ORE DEPOSIT MODELS

RECENT INNOVATIONS, INTEGRATED METHODS, AND CASE STUDIES

SOCIAL AND ENVIRONMENTAL IMPACTS OF RESOURCE DEVELOPMENT DRIVEN BY THE GLOBAL ECONOMY

VITAL METALS FOR THE NEXT CENTURY: FROM EXPLORATION AND DISCOVERY THROUGH PRODUCTION

Mantle-derived Volatile-rich Fluids Induced HREE Enrichment of Previously Differentiated Precambrian Crust to Form the Round Top Laccolith, Trans-Pecos, Texas

Tristan M. Childress, James R. Kyle
University of Texas at Austin, Austin, TX, USA

The Round Top rhyolitic laccolith hosts economic amounts of REEs, particularly HREEs, and is enriched in a variety of other critical elements and minerals. While Round Top and the greater Sierra Blanca Complex (SBC) have been heavily explored from an economic perspective, their geologic origins are poorly understood. Previous geochemical studies suggest the SBC may be the result of crustal melting of differentiated Precambrian plutons via mantle-derived F- and H_2O-rich fluids. Crustal melting began towards the end of flat-slab subduction of the Farallon plate around 40Ma in the Trans-Pecos region, indicated by regional ignimbrite flareups. Slab roll-back induced inflow and circulation of hot asthenosphere beneath freshly exposed SCLM previously enriched with CO^2, F^-, and H_2O from the under-riding oceanic Farallon plate. This influx of hot asthenosphere would have promoted heating and further metasomatism of the SCLM, generating melts and concentrating volatile-rich fluids. These fluids ascended through the lower crust, reaching the MOHO, where mafic melts and volatile-rich fluids would pond. These melts may stagnate and begin differentiation or continue upward via vertical extension-induced structures. Similarly, the volatile-rich fluids may interact with mafic melts or Precambrian crust and, in combination with ductile extension and crustal heating, form local volatile-enriched felsic melts within the MASH zone. These newly formed volatile-enriched melts could then further evolve to enrich the HSFEs and LILEs via F^- ligands and typical differentiation processes, and continue their ascent to the surface via previously emplaced igneous structures such as dikes and sills and or deep-seated NW-SE trending structures derived from crustal compression or extension. After several partial crystallization and eruption events, the later evolved melts erupted to form the Round Top and Little Round Top ore deposits. Regional fluorite deposits and porphyry-style Ag-Pb-Zn mineralization in Texas and Mexico may be genetically related via a deeply rooted magmatic-hydrothermal system.

Geochemistry and Critical Metal Enrichment of the Dinapigue Nickel Laterite Deposit, Isabela, Philippines

Marco Alfredo Barrientos[1], Jillian Aira Gabo-Ratio[1], Julius Pasco[1], Romulo Subong[2]
1. Rushurgent Working Group, National Institute of Geological Sciences, College of Science, University of the Philippines Diliman, Quezon City, Philippines, 2. Dinapigue Mining Corporation, Taguig, Philippines

Nickel laterite deposits are weathering products of mafic and ultramafic protoliths. In the Philippines, most of the laterite deposits form from the weathering of ophiolitic units. These deposits are commonly extracted for their Ni and Fe content, but more recent studies have looked into their potential to hold significant concentrations of critical metals such as Sc, Co, and rare earth elements (REEs). The Dinapigue nickel laterite deposit developed on top of the Isabela Ophiolite, an ophiolite with a mid-ocean ridge (MOR) signature. This is relatively unique with respect to the other studied laterite deposits in the Philippines which all developed from supra-subduction zone (SSZ) ophiolites.

The deposit is underlain by two types of peridotites: dunites and lherzolites. Laterite and bedrock samples were collected from drillcores and mining benches and subjected to petrogaphy and whole rock geochemistry (XRF and ICPMS) analyses. The observed laterite profiles are divided into two parts: an upper, reddish brown, oxide-rich limonite layer; and a lower, greenish, silicate-rich saprolite layer. The results obtained show typical major element enrichment patterns for nickel laterite deposits. Elevated Fe and Al values (up to 75.46 and 8.63 wt.%, respectively) were observed in the limonite layer while Mg and Si are enriched (up to 33.06 and 40.29 wt.% respectively) in the saprolite. Ni tends to have higher concentrations in the saprolite (up to 2.4 wt%). Sc has a positive correlation with both Fe and Al and is more concentrated in the limonite. A field exposure of a Mn nodule was also subjected to bulk geochemical analysis and showed significantly higher concentrations of Co (1.4 wt%), and REEs, particularly Ce (92 ppm) and La (30 ppm). These results indicate that the Dinapigue nickel laterite deposit contains relatively significant amounts of critical metals.

Cobalt and Copper Mineralisation in the Fungurume 88 Deposit

Ryan C. Rosenfels[1], Bjorn P. von der Heyden[1], Jeffrey Dick[2], Kristina Lilova[3], Alexandra Navrotsky[3], Tamilarasan Subramani[3], Brian Woodfield[4], Alexis Gibson[4]
1. Department of Earth Sciences, Stellenbosch University, Stellenbosch, South Africa, 2. School of Geosciences and Info-physics, Central South University, Changsha, China, 3. Arizona State University, Tempe, AZ, USA, 4. Brigham Young University, Provo, UT, USA

The Fungurume 88 deposit is located in the Tenke-Fungurume mining precinct in the Congolese copperbelt, and hosts high grade Cu (up to 4 wt.%) and Co (up to 3 wt.%) in an anomalous stratigraphic position in the usually-barren SD-1b subunit of the lower Shales Dolomitique Formation of the Mines Subgroup. The unit has experienced at least two episodes of fluid flow, giving rise to early stratabound styles of mineralisation (disseminated, stratiform and 'jack' veins) and later cross-cutting veins. The separate timing of the latter style of mineralisation is discerned both by macroscopic field observations and by trace element signatures measured in the contained sulphides, which comprise predominantly carrollite, chalcopyrite and bornite. Detailed optical and electron microscopy work, combined with bulk rock chemistry investigations, revealed that the host rock (a dolomitic shale/mudstone) represents a local facies variation relative to SD-1b units elsewhere in the copperbelt, which are typically coarser-grained dolomitic siltstones or sandstones. This local variant has the required physical and chemical properties to represent an ideal trapping environment for sulphide mineralisation, and its identification opens up exploration opportunities for similar fine-grained SD-1b subunits formed in sub-basins experiencing lower energy regimes during deposition.

The analyses further revealed a decoupling and zonation of the Cu and Co tenors within the SD-1b subunit of the Fungurume 88 deposit. These trends were interrogated further using in situ mineral chemistry analyses and thermodynamic modelling in the software CHNOSZ, coupled with new thermodynamic data for the dominant Co mineral carrollite. These results suggest that pulsed fluid flow may account for the zonation and paragenetic sequencing observed in the Fungurume 88 deposit, and the new thermodynamic data is expected to greatly improve our capacity to understand Cu and Co mineralizing systematics in the greater Copperbelt as well as elsewhere in the world.

Resources for Carbon Neutrality: What Are They, Are They Available, and Where Will They Come From?

Adam C. Simon
University of Michigan, Ann Arbor, MI, USA

What does carbon neutrality mean? What does it require? Is it possible? Here, I will contextualize carbon neutrality through the lens of the metal resources required to transition from a global energy infrastructure dependent on coal, natural gas, and oil, to one entirely reliant on a combination of photovoltaic solar, wind turbines and battery storage. Manufacturing and deploying those renewable energy resources requires dozens of metals, including copper, lithium, nickel, tellurium, cobalt, indium, tin, chromium, and many others. What types of mineral deposits do those metals come from? What are the geologic constraints on their availability? What are the economic constraints on their availability? What are the environmental permitting constraints on the timeframe for production and delivery to market? What are the political constraints on their availability? Analysis of the current mining supply of most metals critical for renewable energy, and the anticipated production from permitted mines at various stages of development, indicate that the supply of most critical metals cannot meet the demand to achieve the climate neutrality goals established by most countries. For example, the supply/demand ratio through the year 2030 for all the following metals is less than one: manganese, chromium, zinc, silver, silicon, molybdenum, rare earth elements (REEs), platinum group metals (PGEs), copper, lithium, nickel, vanadium, cobalt, and graphite. Such data demonstrate significant supply challenges for the transition to a carbon neutral infrastructure within the timeframe the Intergovernmental Panel on Climate Change (IPCC) indicates is necessary to avert average global temperature increasing by more than two degrees Celsius. Further, it takes an average of sixteen years to go from discovery to production. These supply-side facts have major implications for the transition to a post-carbon world and must be addressed by a mindset change and bold leadership at all scales that supports mineral exploration.

An Integrated Perspective on Lithium Mineral Systems

Edward Bunker, Rebecca Bolton, Mark Broadley, Rob Crossley
CGG, Crawley, United Kingdom

Economic concentrations of lithium occur within a range of geological settings and are found within mineral and brine deposits, posing unique challenges and opportunities to geologists in exploring for this critical metal. Whilst over 65% of known lithium resources reside within brines, this is likely an underestimate, as exploration and discovery of unconventional lithium brine resources, such as granite- and basin-hosted brines, is in a phase of rapid growth. This evolution has been driven by surging demand for lithium in batteries, alongside innovations both within direct lithium extraction technologies and geological understanding of the lithium mineral system. Considering these developments, relatively low-grade, high-volume and deep-seated, high-grade resources are being evaluated, and lithium brines are being explored for in new environments, such as oilfields and paleo-salar systems. Furthermore, lithium recovery from brines may be possible in conjunction with production of geothermal energy, potentially offsetting the carbon footprints and economic costs of both.

In order to conduct effective lithium exploration, an overarching understanding of the interlinkage of mineral and brine-hosted lithium systems is essential. By approaching lithium exploration as a mineral systems problem, whereby the key sources, pathways, concentration mechanisms and sinks are considered in detail, it is possible to generate new insights that vector to ore. The presentation will deliver an overview of the lithium mineral system; key examples of applied lithium exploration that focus primarily on lithium brine systems, but leverage fundamental understanding of hard rock geology and lithium sources, will be covered. These case-studies utilise a range of datasets from global- to deposit-scale, and will be utilised to summarise the methodology, data and insights required to generate novel exploration targets.

Exploration 2.0: Systems to Discovery

Nicole Januszczak, Elizabeth Sharman
BHP, Toronto, ON, Canada

The majority of nickel sulphide deposits have been found where portions of the intrusion outcrop at or near surface and the bulk of research on nickel sulphide deposits has focused on geochemical processes in the upper crust where ore accumulates and concentrates. Future discoveries under cover will require an improved understanding of the controls on the evolution of nickel mineral systems. Individual ore deposits are part of much larger mineral systems that are the expression of multiple geological processes. This is useful because we can predict and image systems at larger scales where we cannot possibly hope to detect individual ore deposits. Predictive modelling is currently being tempered by our ability to image critical processes and the four-dimensional evolution of nickel mineral systems. This has highlighted critical gaps and driven an accelerated research program that recognizes the importance of geodynamics, the deep origins of the mineral system, our incomplete understanding of the mineral system and our current biases and assumptions. New concepts are being explored, new datasets acquired, and existing datasets improved or used in new ways, to enable reliable prediction of mineral systems and the detection of ore deposits. Systems thinking is the foundation of the exploration process, and not simply a framework used for prospectivity maps. Conceptual predictive models are tested by exploration work programs and a learning loop is established to ensure our understanding of the system, what we are looking for (proxies), and our exploration toolkits are all optimized. The opportunity in this space is truly extraordinary and requires a capacity uplift in exploration teams and their understanding of how nickel mineral systems form so that they are empowered to plan and execute effective work programs, make decisions and progress projects efficiently through the exploration pipeline.

Critical Minerals in Porphyry Cu-Mo and other Intrusion-related Deposits of the Western United States

Peter Vikre[1], David John[2], Niki Wintzer[3], Fleetwood Koutz[4], Frederick Graybeal[5], Chris Dail[6], David Annis[7]
1. U.S. Geological Survey, Reno, NV, USA, 2. U.S. Geological Survey, Menlo Park, CA, USA, 3. U.S. Geological Survey, Spokane, WA, USA, 4. Consultant, Reno, NV, USA, 5. Consultant, Chatham, NJ, USA, 6. Perpetua Resources, Boise, ID, USA, 7. U.S. Forest Service, Vallejo, CA, USA

The U.S. reliance on imported mineral commodities, manifested during the world and cold war eras, has expanded to include 50 largely imported "critical" minerals (CM) deemed vulnerable to supply disruption. In conterminous western states, numerous chalcophile, lithophile, and siderophile CM are variably concentrated in porphyry Cu-(Mo) (PCD) and other intrusion-related deposits.

Production and reserves of Co, Ni, PGE, Sn, and V in these deposits are mostly insignificant to small, equivalent to months or several years of recent annual domestic consumption (2016-2020). Significant inventories (≥ 2 years of consumption) of Al, Sb, potash, and W in unmined skarn, replacement, vein, and lithocap deposits are ~2 to >8 years of consumption, whereas decades of consumption of As, Bi, fluorite, Ga, Ge, and In exist in these deposit types. Based on concentrations of CM in reserves and captive refinery records, the largest inventories of Sb, As, Bi, Re, and Te, and possibly Ga, Ge, In, Sn, Ti, W, and Zr, are in PCD in Alaska, Idaho, Utah, Arizona, and Montana, and in interim products of refineries that processed PCD for recovery of Cu and Mo, and episodically Re and Te.

Recovery of Sb, As, Bi, Re, and Te, and possibly other CM during concentration and refining of Cu and Mo minerals in ores of operating mines and in unmined reserves is a near-term strategy for reduction or elimination of import reliance. CM, often undetectable in ore, are concentrated (hundreds of ppm to percent) in electrorefining slimes and concentrator tailings, and other than Re and Te, have not been routinely recovered. Because of the large masses of ore processed annually and reserves (hundreds of Mt to Gt), inventories of CM in PCD represent decades to centuries of consumption. However, without recovery processes and mining plans, these inventories do not constitute consumable supplies.

SEG 2022 Conference: Minerals For Our Future

A Workflow to Assess the Critical Mineral Potential of Tailings: Case Study from the Red Dog Mine in Northwest Alaska, USA

Nina Zaronikola[1], Joseph Trouba[1], Elizabeth A. Holley[1], Priscilla P. Nelson[1], Erik Spiller[1], Robin Bullock[2], James Frost[3], Michael Gonzales[3], Lance Miller[4]
1. Colorado School of Mines, Golden, CO, USA, 2. Montana Technological University, Butte, MT, USA, 3. Teck Resources Limited, Anchorage, AK, USA, 4. NANA Regional Corporation, Kotzebue, AK, USA

The elements Co, Ga, Ge, In, Te and V are essential for green energy technologies and are among others characterized as "critical minerals" by the USGS due to their widespread application in global and domestic industries, relatively low abundance, geographic dispersion, and the potential for supply shortfalls. The goal of our project is to identify new approaches for responsible supply of these commodities.

We conducted an assessment of critical mineral recovery potential at the Red Dog Mine, located in the DeLong Mountains, northwestern Alaska. In 2021, Red Dog produced 607,000 tonnes of Zn contained in concentrate. The Red Dog deposit is a black shale-hosted, sedimentary exhalative Zn-Pb deposit (SEDEX) deposit. It is suggested that trace elements, like Ga, Ge, and In along with other critical elements often found in association with zinc could occur as impurities in sulfide minerals (e.g., sphalerite) in various type deposits, including SEDEX deposits.

Here we present an interdisciplinary workflow that we have developed to assess the potential for critical mineral recovery from tailings, using Red Dog as a case study. First, we obtained materials indicative of Red Dog Mine's tailings. This was followed by mineralogical and chemical characterization of the tailings, using optical microscopy, SEM, ICP-OES, and LA-ICP-MS elemental mapping to determine critical element content and their host mineral phases. Based on mineralogical characterization, we performed a metallurgical study investigating potential physical separation and hydrometallurgical processes for critical element recovery and enrichment. New fundamental knowledge generated from such workflows will help predict the economic recovery of critical minerals from tailings at sites worldwide.

Geology and Mineral Resources of the Hicks Dome Fluorite - Barite - Rare Earth Elements Deposit, Hardin County, Illinois

Laurence M. Nuelle, Alan W. Glaser, Eric Livingston
Hicks Dome, LLC, Lutherville, MD, USA

Hicks Dome, a deep-seated structural feature in the Illinois-Kentucky fluorspar district, is host to a fluorite-barite deposit with appreciable amounts of beryllium, niobium, titanium, yttrium, and heavy rare earth elements; all considered critical minerals and elements by the U. S. Geological Survey. Devonian rocks at the dome's apex were uplifted 4,000 feet. Permian alkalic dikes and lapilli-bearing diatremes are indicative of an inferred igneous body at depth. Ubiquitous breccia bodies, 280 to over 2,200 feet in thickness exhibit graded bedding and flow lineation in places. Some intercepts have fragments from overlying stratigraphic units, however those units are relatively unbroken suggesting a lateral component of breccia emplacement. Rock-flour-matrix breccia hosts most resources at depths of 1,500 to 2,000 feet near the dome's apex and depths of 2,500 to 3,000 feet deep away from the apex. A fluorite cut-off grade of 10% (using a 300 feet search radius from drill-holes) yields a geologically inferred resource of 65.8 million tons of 15.8% fluorite, 4.8% barite, 0.16% BeO, 0.2% Nb_2O_5, 0.3% rare earth plus yttrium oxide, and 1.3% TiO_2. A higher cutoff grade of 15% fluorite yields a 24 million ton resource increasing the grade of fluorite to 22.6%, and barite to 7.5%; other metals only slightly increase in grade. The 65.8 million ton resource is greater than the 35 to 40 million tons of fluorite produced from the entire 45-mile-long and 30-mile-wide district. The 24 million ton resource is larger than the largest 10 to 12 million ton deposits in the district. Resources are open-ended and could be increased with additional drilling. Access to drill core has been granted to the U. S. Geological Survey and the Illinois State Geological Survey so that their studies may better define the origin of the deposit, HREE enrichment processes, and the resources it may contain.

Structural Controls and Paragenesis of Mineralization at the Juomasuo Gold-Cobalt deposit, Kuusamo Schist Belt, northeastern Finland

Aaron J. Davies[1, 2], Steffen G. Hagemann[3], Jeffrey Foster[1], Walter Witt[4], Francois Voute[5]
1. Latitude 66 Cobalt Oy, Oulu, Finland, 2. Aztec Geoscience and Exploration Ltd, Bude, United Kingdom, 3. Centre for Exploration Targeting, University of Western Australia, Nedlands, WA, Australia, 4. The Walter Witt Experience, South Perth, WA, Australia, 5. Francois Voute Geological Services, Fremantle, WA, Australia

The Juomasuo Au-Co deposit is the largest known mineral deposit in the Paleoproterozoic Kuusamo Schist Belt. It is situated at a prominent flexure within the ~30km long, NE-SW-striking Käylä-Konttiaho Antiform, a structural corridor that hosts several Au-Co-Cu deposits. The host rocks comprise a highly deformed, upper greenschist metamorphosed package of sedimentary, mafic volcanic and mafic-ultramafic rocks.

Five major deformation events (D_{1-5}) are recorded in the host rocks of Juomasuo. Earliest deformation (D_1) is represented by poorly defined folds (F_1) and faults, and V_1 veins containing albite±quartz±carbonate. The Käylä-Konttiaho Antiform is interpreted to have developed during D_1.

The D_2 event was multi-staged and resulted in an early, regional penetrative S_2 foliation and tight F_2 folds and late brittle-ductile shear zones. The major Au-Co mineralization event of the Juomasuo deposit occurred during the late stages of D_2 and is mainly associated with NW-striking, SW-dipping shear zones. Veins are rare in ore zones but, where observed comprise quartz-chlorite-pyrrhotite-pyrite±muscovite (V_2) and form crack-seal types in competent rocks or deformed sigmoidal or boudinaged types in shear zones. Peak metamorphism coincides with D_2.

D_3 structures include NE-SW-trending open folds, S_3 foliation, NW- to SE-dipping faults and V_3 veinlets containing pyrite-carbonate. Minor Au mineralization took place during D_3.

Seven distinct hydrothermal alteration assemblages are recognized throughout the deposit and are linked to the deformation events and associated structures. Early, pre-ore formation stages resulted in Na-(Ca), Mg-Fe-Ca and Mg-Fe-K metasomatism. Intense albite alteration of the Na-(Ca) stage created extreme competency contrasts within the stratigraphy, which localised deformation and subsequent hydrothermal fluid flow during D_2.

Hydrothermal alteration assemblages of the late D_2 major Au-Co mineralization event are divided into: chlorite-pyrrhotite-biotite-quartz-apatite-rutile-chalcopyrite-pyrite±gold assemblages, and muscovite-quartz-pyrite-pyrrhotite-apatite-rutile-cobaltite-linnaeite-cobaltpentlandite-gold assemblages.

Characteristic structural, hydrothermal alteration and mineralization features of Juomasuo are interpreted to be similar to those of metasedimentary rock-hosted cobalt deposits of Blackbird district, Idaho.

Critical Mineral Potential of Jurassic Magmatic Systems in the Eastern Basin and Range

Stephanie E. Mills
Utah Geological Survey, Salt Lake City, UT, USA

The intrusive activity of the Jurassic era in the Basin and Range Province is well documented (e.g., Barton et al., 2011) and associated mineralization is generally less significant than later magmatic events, leading to a lack of interest in Jurassic-age intrusions as an exploration target. However, recent discovery of a shallowly buried, strongly altered and weakly mineralized Jurassic magmatic system in Utah (Thompson Knolls) has revived interest in the mineral potential of intrusions in that time period. Jurassic intrusions are known to generate a range of mineral systems, some of which can be high grade and/or contain considerable critical mineral potential, such as at the Gold Hill mining district, Utah. Gold Hill also demonstrates the influence of Jurassic intrusions on later magmatic events by serving as pre-existing pathways of crustal weakness focusing subsequent magmatic emplacement. As noted, later magmatic events tend to have more significant mineral potential, and Gold Hill provides a blueprint for interaction between a Jurassic pluton and later Eocene stock, highlighting the need to understand ages of all magmatic activity in blind exploration targets. Given the diverse mineral potential of Jurassic intrusive systems and their ability to influence later magmatic events, an evaluation of the exploration potential and approach to buried Jurassic systems is discussed, with an emphasis on how these systems factor into the critical mineral exploration scene.

Barton, M. et al., 2011, Jurassic igneous-related metallogeny of SW North America: GSN Symposia Proceedings.

Metallogeny and Exploration Strategy for Alkaline Volcanic Rocks Hosting World Class Be-U-F Mineralization at Spor Mountain, Utah, U.S.A.

Nora K. Foley[1], Robert A. Ayuso[1], Jorge A. Vazquez[2], John A. Jackson[1], Rani Indela[1], Damon Bickerstaff[1]
1. US Geological Survey, Reston, VA, USA, 2. US Geological Survey, Menlo Park, CA, USA

New geochemical data for the Spor Mountain Formation (SMF) shed light on the metallogeny of rocks hosting the largest known deposit of beryllium in the world. The deposit and hostrocks (a lower Be-U-F-rich tuff and upper topaz-rich rhyolite) are part of an extensive sequence of Miocene-to-Oligocene alkalic rhyolitic tuffs and lavas occurring near the boundary between thicker and thinner crust. A broader understanding of the genesis of SMF is critical for evaluating regional and global exploration strategies for volcanic-hosted Be.

Key attributes of SMF include a magmatic Be-U-F enrichment stage from melting of continental lithosphere and later hydrothermal/metasomatic Be-enrichment stages. The magmatic stage is characterized by distinctive elemental (Be, REE, Y, Ti, by SHRIMP) contents in zircon. In post-magmatic stages, F-rich fluids reacted with tuff, altering glass and carbonate detritus to clay-rich sequences (primarily montmorillonite) containing high-grade (bertrandite + fluorite + opal) mineralization. The SMF shows contrasting behavior of Nb and Ta reflecting effects of prolonged fractional crystallization (e.g., large Eu anomalies) and pervasive hydrothermal alteration by F-rich aqueous fluids. A shared source and evolution for rhyolite and tuff are indicated by (1) regular, stacked subparallel, chondrite-normalized REE patterns, (2) overlapping sub-vertical arrays for Pb isotopes ($^{207}Pb/^{204}Pb$ ~15.60-15.70 at $^{206}Pb/^{204}Pb$ ~18.60), and (3) negative values of e_{Nd} (average ~-8.0). Pb isotopic ratios for SMF do not match slopes, initial isotope values, nor basement ages for precious/base-metal districts in Utah. Crustal ages from Nd isotopes (average T_{DM} ~2.2 Ga) are older than basement ages obtained by Pb isotopes. Rare-metal indices of granite-related mineralization (e.g., Nb/Ta, Zr/Hf) show SMF values overlap fields for globally-distributed Be-bearing granitic/rhyolitic deposits, for example, the Ermakovka F-Be deposit (Russia) and Baiyanghe Be-U deposit (China); however, none of these deposits match the mineralogy, geologic setting, or rare metal chemistry of the Spor Mountain system.

Manganese Deposits within the Cape Supergroup and their Relation to Thermal Springs

Rutger La Cock, Bjorn von der Heyden, Jared Van Rooyen, Tersia Conradie, Daniel Ferreira
Stellenbosch University, Cape Town, South Africa

The mechanisms controlling the formation of manganese deposits within the ~300 Ma Cape Supergroup (South Africa) are not well constrained. Manganese is a critical element not only in steelmaking, but in the production of batteries which are vital to the rapidly growing electric vehicle industry. Much work has been done on the mineralisation seen in the western portions of the Cape Fold Belt (CFB), however the same is not true for that of the central and eastern portions of the CFB. This study looks at the CFB in its entirety and serves to determine the mode of precipitation of manganese. In an effort to reach this goal, a model involving thermal springs was investigated with springs being described in terms of their physiochemical, chemical, isotopic and microbiological characteristics. Two Mn deposit classes exist within the CFB: vein hosted, cryptomelane-goethite dominated mineralisation and pyrolusite-goethite surficial deposits intimately associated with thermal spring activity. Manganiferous springs are shown to have significantly higher salinities (> 95 ppm) than non-manganiferous springs (< 70 ppm) and this is intrinsically linked to manganese concentration as chloride acts as a transport ligand for Mn. PHREEQC modelling software was utilized to show that mixing between warm, fossil spring water and cool, modern groundwater is a viable means for the precipitation of manganese oxides. Microbial communities were classified to determine whether biomineralization may have contributed to oxide precipitation. The presence of multiple different species of manganese oxidising bacteria indicates that this process is biologically-mediated. Burkholderiales bacteria which are associated with manganiferous springs was identified as a biomarker and has potential to be used as an exploration tool. This work shows that there is an intimate relationship between thermal springs and manganese mineralisation within the CFB, and has implications for our understanding of vein-hosted and surficial Mn deposits located elsewhere globally.

Sulfur Isotope and PGE Decoupling During Silicate Melt-Sulfide Liquid Equilibration

Anne B. Virnes[1], Marco L. Fiorentini[1], Stephen J. Barnes[2], Stefano Caruso[2, 1], Laure Martin[3], Matvei Aleshin[3], Louise Schoneveld[2], Malcolm Roberts[3], Quentin Masurel[1], Nicolas Thebaud[1]
1. Center For Exploration Targeting, School of Earth Sciences, The University of Western Australia, Crawley, WA, Australia, 2. CSIRO Mineral Resources, Kensington, WA, Australia, 3. Centre for Microscopy, Characterisation and Analysis, The University of Western Australia, Crawley, WA, Australia

Komatiites are inherently sulfide-undersaturated and require external sulfur from country rocks to generate immiscible sulfide liquid, which concentrates metals to form economic nickel-sulfides. Recent work has shown that mass-independent fractionation of S-isotopes (MIF-S, denoted $\Delta^{33}S$) may inform on the nature of external sulfur sources. However, it is possible that the MIF-S signature of S sources may be diluted by S exchange between the assimilated sulfide xenomelt and the komatiite silicate melt. To address this, we investigated the effect of silicate melt-sulfide liquid equilibration (R-factor) on the resulting MIF-S signatures of pentlandite-rich ore from the Mount Keith MKD5 Ni-deposit, Agnew-Wiluna Greenstone Belt, Western Australia.

We carried out in-situ SIMS multiple sulfur isotope analysis on pentlandite from a suite of sulfide-bearing samples to assess the MIF-S relationship with R-factor at the scale of individual sulfide droplets that were also analyzed for Pd using in-situ LA-ICP-MS. The results show that in-situ Pd in pentlandite correlates poorly with Pd tenors derived from whole-rock S and Pd assays, due to the occurrence of Pd-bearing minerals.

The observed variability in Pd-tenor and MIF-S signature suggests that the latter is not controlled by R-factor. Rather, the spread of MIF-S signatures suggests that the sulfide xenomelt was initially heterogeneous and that MIF-S did not fully equilibrate and homogenize. This suggests that the chemical equilibration of S-isotopes is decoupled from the equilibration of platinum group elements in a komatiite melt. Our work shows that even in the hottest, most dynamic, and likely fastest equilibrating magmatic systems, xenomelt sulfides may still preserve to some extent their initial sulfur isotopic compositions, reflecting the wide range of crustal sulfur reservoirs that were available during komatiite emplacement. Decoupling of MIF-S and Pd tenor may be the result of relatively slow self-diffusion of S relative to the diffusion of Pd in silicate melt.

Magmatic and Hydrothermal Controls on the Evolution of Ni-Co Mineralization in the Late Jurassic Alaskan-type Mafic-Ultramafic Turnagain Complex

Kiera Broda, Anthony Williams-Jones
Mcgill University, Montreal, QC, Canada

The Turnagain Intrusion (currently being explored and evaluated by Giga Metals Corp.) is a Late Jurassic Alaskan-type mafic-ultramafic complex hosting magmatic nickel-cobalt sulfide (pyrrhotite-pentlandite) mineralization. Alaskan-type magmas are typically undersaturated in respect to sulfur and require its addition to reach saturation and form an immiscible sulfide liquid. Values of $\delta^{34}S$ varying from the magmatic range of 0 ± 3 to -17.9 ‰ are consistent with a hypothesis that the additional sulfur was supplied by assimilation of pyrite-bearing graphitic phyllite of the Middle to Late Mississippian Road River Formation.

The sulfides are disseminated to net textured and the pentlandite occurs as exsolution lamellae and more commonly as discrete crystals or as large masses within pyrrhotite, which may indicate peritectic crystallization of the pentlandite from monosulfide solid solution and sulfide liquid. However, as the host rocks have been variably serpentinized and crystals of pentlandite are observed in serpentine veins, an alternative explanation is that iron (and sulfur) was remobilized by the serpentinizing fluids and that the much less mobile nickel was immobilized and recrystallized as pentlandite. It is also possible that nickel was released from olivine (it contains up to 0.4 wt.% Ni) during serpentinization and sequestered by sulfur released from pyrrhotite to form pentlandite. These processes could help explain the poor correlation of nickel and sulfur contents in the ores that has provided a challenge in developing a robust metallurgical protocol and was the motivation for this study. Future work will test these hypotheses by evaluating the roles of magmatic and hydrothermal processes in producing this unusual nickel-cobalt deposit.

The Thermodynamic Properties of Carrollite (CuCo2S4) and Their Application in Modeling Hydrothermal Co(-Cu) Ore-formation

Robert Collar, Kirsten Rempel, A. E. Williams-Jones
Department of Earth and Planetary Sciences, McGill University, Montreal, QC, Canada

The production of electric vehicles has fueled a strong demand for Co, a critical component of the batteries used by these vehicles, and is projected to drive Co demand for the foreseeable future. Recently, much of this demand has been accommodated by increased production in the Central African Copperbelt (CACB), particularly in the Democratic Republic of the Congo (DRC), which accounts for ~70% of global Co production (2021). In the absence of robust Co recycling programs, increased exploration in this region and elsewhere will be required to meet future demand. However, the processes responsible for hydrothermal Co mineral deposition, the main means of Co concentration, are poorly understood, hampering exploration for this critical metal. A major reason for this poor understanding is that there are few thermodynamic data for Co minerals and none for carrollite ($CuCo_2S_4$), the principal ore mineral of Co in sediment-hosted deposits of the type exploited in the CACB. Consequently, it has not been possible to model the physicochemical conditions of hydrothermal Co transport and deposition and develop quantitative models of oregenesis on which to base exploration strategies.

In this study, calorimetry and solubility measurements are used to determine values of the standard thermodynamic parameters ($\Delta_f G°_{Tr,Pr}$, $\Delta_f H°_{Tr,Pr}$, $S°_{Tr,Pr}$, $C°p$, and $V°_{Tr,Pr}$) for carrollite at reference conditions (i.e., T_r = 25 °C and P_r = 1 bar). A value for $V°_{Tr,Pr}$ was calculated from published unit cell parameters. Low-temperature calorimetry measurements of $C°_p$ have been made in order to calculate $S°_{Tr,Pr}$. These data, in conjunction with solubility data collected for carrollite in equilibrium with an aqueous NaCl-HCl solution at 200 °C and saturated water vapor pressure, permit calculation of $\Delta_f H°_{Tr,Pr}$ and $\Delta_f G°_{Tr,Pr}$. Using the Python libraries Reaktoro and ThermoFun, we employ these new data for carrollite to build plausible geochemical models for the formation of sediment-hosted Cu-Co deposits.

Potential Domestic Sources of Critical Minerals for Our Energy Future

Jane M. Hammarstrom[1], Laurel Woodruff[1], Connie Dicken[1], Warren Day[2]
1. USGS, Reston, VA, USA, 2. USGS, Denver, CO, USA

Critical minerals needed for the U.S. transition to a clean energy economy include cobalt, lithium, nickel, and graphite. These commodities were historically produced on small scales in the U.S., often during wartime, but now are primarily sourced from other countries. As part of the Earth Mapping Resources Initiative (Earth MRI) project, the USGS, in partnership with State geological surveys, identified broad areas of the country (focus areas) that are known to, or have potential to, host deposits containing these and other critical minerals. These focus areas are based on recognition of permissive geologic settings and relevant mineral occurrences considered in a mineral system/deposit type framework. GIS representations of focus areas and supporting information compiled from existing data are the starting point for prioritizing areas across the country for acquisition of new geologic mapping, geochemical data, airborne geophysics, and lidar data. For example, the IOCG system in the Idaho Cobalt Belt is one focus area where Earth MRI is funding new mapping efforts and a geophysical survey across the belt. Cobalt also occurs as a byproduct in some sediment-hosted copper and MVT deposits in basin brine systems, with nickel in mafic magmatic systems, and in laterites formed in chemical weathering systems. Lithium resources in Nevada occur in residual brines and clays in lacustrine evaporite systems, and new mapping is underway in the eastern U.S. highlighting spodumene [$LiAl(SiO_3)_2$] in LCT pegmatites. Two new airborne electromagnetic surveys will refine permissive areas for graphite deposits in Alaska and Alabama. Data from projects such as these will provide new information for future mineral resource assessments and facilitate exploration and development planning by State partners, private industry, and land use managers. A major challenge will be identification of economically viable deposits and determination of their suitability for development and recovery of critical minerals.

Cornish Lithium: Unlocking Unconventional Lithium Deposits to Accelerate the Energy Transition in the UK

Rebecca Paisley, Chris Yeomans, Lucy Crane
Cornish Lithium, Penryn, United Kingdom

The energy transition is focused on rapidly decarbonising global energy consumption in a bid to meet the goals of the 2015 Paris Agreement. Simultaneously, there is a shift towards geopolitically secure and responsible supply chains, leading to renewed interest in mineral exploration and refining in Europe. One key metal for the transition is lithium for use in batteries, presently produced from South American salars or Australian spodumene resources. Existing resources must increase production. However, to meet the expected demand driven by electrification of the automotive industry, novel and unconventional resources must be discovered and brought online. The result is a revolution in new exploration and processing techniques as the mining industry races to keep up with future demand for critical metals.

In Cornwall, UK, lithium concentrations were first reported in geothermal fluids in 1864 and the county is underlain by one of the world's largest lithium-enriched granitic batholiths. Cornish Lithium Ltd is an innovative mineral exploration company, focused on the responsible extraction of lithium from both sources. Production from hard-rock is centred on a non-operational kaolin pit. In 2021, a JORC compliant Maiden Inferred Resource Estimate was announced with 51.7 Mt at 0.11% Li. By utilising existing infrastructure and Lepidico's L-MAX®/LOH-MAX® hydrometallurgical process, which does not use a high temperature roast, production is expected to have a considerably reduced environmental impact with respect to current hard-rock sources of lithium over a 20-year life-of-mine. Lithium extraction from geothermal waters provides Cornwall an opportunity to produce low-carbon lithium, heat and other metalliferous co-products to revolutionise industries across the region. Exploration boreholes have confirmed the presence of "globally significant concentrations of lithium" in geothermal waters extracted from natural fracture zones. Together, these Cornish Lithium projects show how innovation can lead to renewed exploration and production of critical metals in historic mining districts.

Trace Element Signatures of Quartz from Li-Cs-Ta Pegmatites as Exploration Vectors for Li-mineralization: Examples from Austria and Ireland

William Keyser[1], Axel Müller[1], Jasper Berndt[2], Ralf Steiner[3], Tanja Knoll[4], Julian F. Menuge[5]
1. Natural History Museum, University of Oslo, Oslo, Norway, 2. Institut für Mineralogie, Westfälische Wilhelms-Universität Münster, Münster, Germany, 3. Geo-Unterweissacher GmbH, Hochfilzen, Austria, 4. Geological Survey of Austria, Vienna, Austria, 5. Irish Centre for Research in Applied Geosciences (iCRAG), University College Dublin, Dublin, Ireland

Increased demand for lithium has led to renewed interest in known lithium-cesium-tantalum (LCT) pegmatite occurrences and an exploration surge for new lithium resources, particularly in regions with low raw material production and heavy importation (e.g., the European Union). Trace element concentrations in quartz were determined from two well-known LCT pegmatite regions in Europe to test the applicability of quartz as a pathfinder mineral for Li-mineralization. The Moylisha area (SE Ireland) and the Eastern Alps (Austria) represent two distinctly different geological histories. The spodumene-bearing (mineralized) and spodumene-free (barren) pegmatites at Moylisha form a late Silurian to early Devonian NE-SW-striking pegmatite belt that intruded the East Carlow Deformation Zone along the SE margin of the late Caledonian S-type Leinster Batholith. Contrastingly, pegmatites of the Eastern Alps are associated with late Permian lithospheric thinning/basaltic underplating that formed a larger pegmatite province spanning the Eastern Alps. They were metamorphosed at up to eclogite-facies conditions during the Alpine orogeny.

All analyzed pegmatites contain assemblages including feldspar, quartz, muscovite ± spodumene. LA-ICP-MS analysis shows quartz from mineralized pegmatites in both regions as distinguishable from that in barren pegmatites by higher concentrations of Al, Li, Ge and B, whereas barren pegmatite quartz contains higher Ti. Principal component analysis of quartz data shows that Al and Li signatures characterize quartz in pegmatites from Moylisha whereas Ge and B signatures characterize quartz in pegmatites from the Eastern Alps. In quartz from pegmatites and associated leucogranites across the Eastern Alps, concentrations of Al, Li, Ge and B increases eastward with highest concentrations occurring at the Wolfsberg lithium deposit, showing its utility as a vector towards significant Li-mineralization. The coupled substitution $Al^{(3+)} + Li^{(+)} \rightarrow Si^{(4+)}$ governs the Li content in quartz, and concentrations of >100 $\mu g g^{-1}$ Al and >30 $\mu g g^{-1}$ Li represent an important threshold for spodumene mineralization.

Low-density Mapping of Lithium Concentration in Stream Sediments and Definition of Geochemical Background in Minas Gerais State, Brazil

Luísa S. Guerra[1], Mariangela G. Leite[1], Maria Augusta G. Fujaco[1], Eduardo D. Marques[2]
1. Ouro Preto Federal University, Ouro Preto, MG, Brazil, 2. Brazilian Geological Survey, Belo Horizonte, MG, Brazil

Lithium is among the chemical elements that represent an emerging world demand, due to its use in high-tech products as lithium-ion batteries. The Investment Partnership Program Council of Brazil – CPPI, in June 2020, listed the characteristics of mineral goods that can qualify a mining investment project with government priority, among them there is lithium. One of the most used investigation methods in mineral prospecting is based on the study of stream sediments, whose prediction and reliability in detecting anomalous concentrations help to map elements availability. Stream sediments, due to their genetic complexity, can be considered representative of the geology upstream the sampling point in hydrographic basins, ideal for low-cost regional mineral exploration. This study aims to: apply different statistical techniques (TIF, MAD) to define geochemical background values of lithium concentration in these stream sediments; in a GIS environment, acquire the distribution and concentrations of lithium in seven different geotectonic units (GU) and identify potential mineral deposits. In partnership with CPRM (Geological Survey of Brazil), 3962 stream sediments were collected at a low-density scale between 2009 and 2012 throughout the state of Minas Gerais. They were treated by the laboratory SGS-GEOSOL® where the samples were dried at 60°C, quartered and sieved in meshes with an opening <80 mesh. The sieved pulps were subjected to spraying and 0.5g of each was mixed with 3ml of aqua regia (HCl + H_2NO_3 + water, in the proportion 3:1:2). The resulting extract was diluted to 1ml of water and subjected in ICP-OES and ICP-MS for 53 elements. Based on uni-varied stats (box plots) and relations to GU, the higher values are in Iron Quadrilateral (Craton part), Araçuai Belt and São Francisco Basin (Craton). According to bi-varied stats, Li has an increasing chemical relation with Al < K < Rb < Be < Cs.

U-Pb Geochronology of LCT and NYF Pegmatite Occurrences of the Northern Appalachians

Christopher Holm-Denoma[1], Laura Pianowski[1], Dwight Bradley[1], Myles Felch[2], Sean Kinney[3], Steven Jaret[4], Amber Whittaker[5], Robert Marvinney[5]
1. U.S. Geological Survey, Denver, CO, USA, 2. Maine Mineral and Gem Museum, Bethel, ME, USA, 3. Lamont Doherty Earth Observatory, Palisades, NY, USA, 4. American Museum of Natural History, New York, NY, USA, 5. Maine Geological Survey, Augusta, ME, USA

Pegmatites host significant critical minerals resources worldwide. Lithium-cesium-tantalum pegmatites (LCTP) are the more abundant of the two highly evolved granitic pegmatite types, along with niobium-yttrium-fluorine pegmatite (NYFP)-type, and are enriched in rubidium, beryllium, boron, and tin. Hypotheses of pegmatite generation include derivation from extreme fractionation of parental granitic magmas, direct anatexis of enriched parent rocks, and formation below salars in orogenic belts.

Pegmatite mineral geochronology is the best way to determine their temporal relations. Zircon geochronology is problematic due to the presence of xenocrysts, inheritance, and Pb-loss/High-U. Cassiterite, the main ore mineral of tin, is common in LCTPs and some NYFPs and is a proven geochronometer. New LA-ICPMS U-Pb geochronology analyses of cassiterite from several LCTPs in western and central Maine are ~350-250 Ma. NYFPs from eastern New Hampshire are ~180 Ma—clearly related to the Mesozoic White Mountain magmatic series (WMMS). Altogether there were at least five episodes of evolved pegmatite genesis in the northern Appalachians.

Strikingly, most LCTP pulses occur at the terminus of magmatic events or during apparent regional magmatic lulls. This suggests that LCTP genesis in western Maine is not related to fractionation of coeval parental granitic bodies but, alternatively, is related to processes that followed major tectonomagmatic events like the Neoacadian and Alleghanian orogenies. That LCTPs follow tectonism is consistent with anatexis as a contributor to the genesis of LCTP. Relatively high-precision age constraints from cassiterite geochronology allows us to correlate paleogeographic positioning during formation. Western Maine at ~260 Ma was at an arid latitude making it feasible that fluids from salars could enrich the pegmatite source.

Cassiterite from select NYFPs in New Hampshire are coeval with emplacement of the WMMS and appears to be directly related to that event—NYFP formation is a fundamentally different process than the formation of LCTPs the region.

SEG 2022 Conference: Minerals For Our Future

A Devonian Re-Os Age for the Bou Azzer Co-Ni-As Deposit, Morocco

Holly Stein[1, 2], Saïd Ilmen[3], Thomas Oberthür[4], Aaron Zimmerman[1], Siri Simonsen[2], Lhou Maacha[3], Mohamed Zouhair[3]
1. AIRIE Program, Colorado State University, Fort Collins, CO, USA, 2. Department of Geosciences, University of Oslo, Oslo, Norway, 3. Bou Azzer Cobalt Mine, Greenfield Cobalt Exploration, Managem Group, Bou Azzer, Morocco, 4. Bundesanstalt für Geowissenschaften und Rohstoffe, Hanover, Germany

The Co-Ni-As deposits of Bou Azzer floored by the Anti-Atlas Supergroup have a rich orogenic history reaching back to the Mesoproterozoic. Relicts of a rift-generated, then obducted oceanic crust hosting the Bou Azzer mineralization present as highly sheared ~760 Ma allochthonous ophiolite complexes intruded by younger Neoproterozoic quartz diorites. Prevailing models advocate Neoproterozoic, Variscan, and/or Late Triassic ore genesis. Although analytically challenging, Re-Os dating of micron-scale molybdenite (MoS_2) intergrown with the cobalt arsenide ore mineral skutterudite provided the first indication for a Devonian mineralization event (400-350 Ma-1).

Here, we present new Re-Os ages derived from a newly discovered, unusual molybdenite occurrence underground. Syn-kinematically deformed with widespread chlorite, molybdenite exhibits polished mirror-like surfaces and woody textures. SEM photos show felt-like masses of poorly crystalline to near amorphous molybdenum disulfide (jordisite). Chalcopyrite is widely present, as are tiny cobaltite (CoAsS) crystals. Mistakenly described as sphalerite, an adamantine yellow to red translucent enstatite (MgSiO3) occurs in its rare pure end-member form. Coffinite, a uranium silicate rarely seen as crystals, forms clusters of spectacular tetragonal crystals. Clausthalite (PbSe) is widely present. Most striking, however, is the widespread presence of significant carbon in both sulfides and silicates. Our results outline two distinct events at 385-390 and 375 Ma for molybdenite-chalcopyrite-cobaltite-skutterudite deposition. The 375 Ma molybdenite-bearing event has significant common Os yielding an isochron with an initial 187Os/188Os suggesting influx of primitive fluids in the Devonian.

Proterozoic shales as source rocks are difficult to reconcile with these Re-Os data. Rather, Devonian mineralization appears to have involved fluids from the mantle and potentially Neoproterozoic oceanic crust. The widespread association of carbon with the mineralizing system has yet to be reconciled, but is critical to the ore deposit model.

This project was initiated and funded by the Managem Group.

1-Oberthür, Stein et al (2009) Economic Geology 104: 1065-1079.

The Fishtie Cu-Co Deposit, NE Zambia: Stratigraphy, Structural Architecture, Hydrothermal Alteration, and Mineralization

Subaru Tsuruoka, Danny Hnatyshin, Murray Hitzman, Koen Torremans
Irish Centre for Research in Applied Geosciences (iCRAG), University College Dublin, Dublin, Ireland

The ores of the Central African Copper Belt (CACB) are hosted in the Neoproterozoic Katangan Supergroup which consists of rift-related sediments (Lower Roan Subgroup), platformal sediments (Upper Roan Subgroup), and shallow marine sediments (Mwasha Subgroup, Nguba Group, and Kundelungu Group). The Fishtie deposit is an important example of Cu-Co mineralization and is located in the south-easternmost portion of the Zambian Copperbelt in the Lusale basin. The Fishtie deposit is characterized by Cu and Co sulfide mineralization hosted in the Nguba Group. Here we provide a full geologic description of the deposit that can be used to help evaluate the nature and evolution of hydrothermal systems at Fishtie and the greater CACB.

The lowermost Nguba Group rocks consist of a <300m-thick succession which directly overlies basement rocks or rocks of the Upper Roan Subgroup; rocks of the Lower Roan Subgroup are absent. The Nguba Group consists of diamictite and siltstones of the Grand Conglomérat which are overlain by carbonaceous siltstones of the Kaponda Formation and the dolomites and carbonaceous siltstones of Kakontwe Formation. The Grand Conglomérat is comprised of five individual debrite units whose thicknesses were controlled by basement topography and synsedimentary normal faults.

An assemblage of bornite-chalcopyrite occurs above positive basement topographic relief and adjacent to synsedimentary normal faults. This sulfide assemblage transitions outboard to chalcopyrite-pyrite and pyrite-sphalerite assemblages. Cobalt mineralization occurs in the eastern portion of the deposit where cobaltite is disseminated in carbonaceous siltstone and also occurs in dolomite-quartz veins. Alteration is consistent throughout the deposit with early biotite and quartz replaced and overgrown by chlorite, dolomite and sulfides. A mean model Re-Os age of 492+/-3 Ma (n=3) from vein-hosted cobaltite indicates that cobalt mineralization followed the main stage of the Lufilian Orogeny.

Isotopic and Trace Element Signatures of Calcite, Apatite and Zircon from Carbonatite Liquid Associated with Cu-Ni-PGE Mineralization

Maria Cherdantseva, Marco Fiorentini
Centre of Exploration Targeting, School of Earth Sciences, The University of Western Australia, Perth, WA, Australia

Based on previous detailed mineralogical, petrological and experimental studies, we put forward a model whereby immiscible volatile-rich carbonatite liquid may play a key role in the transportation of Ni-Cu-PGE-rich sulfides from the mantle to the crust. The crystallized product of carbonatite liquid forming haloes surrounding magmatic sulfide globules generally comprises plagioclase, clinopyroxene, amphibole, phlogopite, carbonate, apatite, ilmenite and zircon. This paragenetic association can be found in many examples of magmatic sulfide deposits globally. To further refine our model, we have investigated and present here isotopic and trace element composition of accessory minerals (calcite, apatite and zircon) from three mineralized intrusions located at variable depths: lower crustal intrusion - Valmaggia pipe, Italy; middle crustal intrusion – Broken Hill, Australia; upper crustal intrusion – Rudniy, NW Mongolia. Results allowed us to confirm the carbonatitic origin of the haloes surrounding the magmatic sulfide globules and link their signatures in REE distributions among different phases to the depth of emplacement of the individual intrusions. The isotopic signatures of sulfur, carbon and oxygen in apatite, calcite and zircon, respectively, support a mantle source origin for the carbonatite liquid in mineralized intrusions emplaced at different depths and within very different country rocks. The proposed model of carbonatite-enhanced physical transport of magmatic sulfides may lead to a reconsideration of existing paradigms on the formation of orthomagmatic systems, enhancing the predictive and detective capability of exploration techniques to target elusive mineralized systems enriched in nickel, copper and the valuable platinum group metals.

Graphite Re-Os Dating

Jonathan Toma[1], Robert A. Creaser[1], Colin Card[2], Richard A. Stern[1], Thomas Chacko[1], Matthew Steele-MacInnis[1]

1. University of Alberta, Edmonton, AB, Canada, 2. Geological Survey of Saskatchewan, Regina, SK, Canada

Natural graphite is one among a handful of critical materials used in emerging technologies, such as lithium-ion batteries. In order for the supply of natural graphite to keep a pace with growing demand new graphite deposits will have to be discovered [1]. However, the search and discovery of such occurrences has hitherto been limited by a lack of graphite chronology information pertinent for understanding ore field relationships. Without this direct information, the timing of graphite mineralizing events have been inferred from other minerals in adjoining host rock materials, which may not be part of the same paragenetic assemblage.

Here we showcase recent developments for age dating natural graphite using the Re-Os decay system [2]. We begin by investigating the Re contents for graphites (n = 17) formed in metamorphic, hydrothermal, and meteoritic environments. In all three cases, graphite Re abundances exceed upper continetal crustal (UCC) Re values but are analagous to Re values found in terrestrial sulfides, organic-rich sedimentary rocks, and hydrocarbons.

A subset of these graphites (n =3) were selected for graphite Re-Os dating. These include two hydrothermal graphite samples formed in lower-crustal shear zones (Wollaston-Mudjatik Transition [Gr_{WMT}], Canada]) and tanzanite-tsavorite gemstone deposits (Merelani Hills [Gr_{MH}], Tanzania) and one metamorphic graphite sample formed in the Franciscan subduction zone (Laytonville Quarry [Gr_{LQ}], USA).

High-precision Re-Os graphite dates were obtained (Gr_{WMT} = 1731.52 ± 7.43 Ma; Gr_{MH} = 586.89 ± 2.39 Ma; Gr_{LQ} = 161 Ma]) from all three samples. These graphite Re-Os dates are consistent with the inferred formation ages of graphite mineralization constrained by other techniques. The success of the graphite Re-Os dating method now allows for the possibility of other graphite deposits to be constrained and their formation models tested within a robust temporal framework.

[1] Olson et al. (2016)

[2] Toma et al. (2022)

Integrating Mineralogical and Geochemical Data Across Scales for Improved Geophysical Data Interpretation: a Case Study from the Idaho Cobalt Belt, USA

Daniel E. Schmidt[1], Katharina Pfaff[1], Albert Hofstra[2], Geoffrey Phelps[2], Reed Lewis[3], Thomas Monecke[1], Warren Day[2]
1. Center for Advanced Subsurface Earth Resource Models, Department of Geology and Geological Engineering, Colorado School of Mines, Golden, CO, USA, 2. U.S. Geological Survey, Denver, CO, USA, 3. Idaho Geological Survey, University of Idaho, Moscow, ID, USA

The Idaho cobalt belt (ICB) includes several stratabound cobalt and copper deposits hosted in metasedimentary rocks of the Belt basin that contain significant cobalt. Despite its economic and strategic importance, the processes that form these deposits are still poorly understood, and therefore the potential for concealed critical mineral resources is uncertain. The U.S. Geological Survey's Earth Mapping Resources Initiative has funded geologic mapping by the Idaho Geological Survey and, in conjunction with industry partners, aeromagnetic and radiometric surveys across the ICB.

The goal of this collaboration is to improve our understanding of the potential for critical mineral resources in the ICB and elsewhere by designing a workflow that combines both mineralogical and geochemical characterization across a range of scales. These fully co-registered heterogeneous remote sensing and sub-surface datasets will help inform geophysical data interpretation and subsequent construction of interpretive 3D geologic models to better understand the mineralogy and geochemistry of the alteration and ore zones.

Drone-based hyperspectral surveys, to be completed in the summer of 2022, will be combined with continuous XRF and hyperspectral core scans, and magnetic susceptibility data collected on drill core and automated mineralogy from select thin sections to form heterogeneous but fully co-registered datasets. Selected drill core from the Iron Creek deposit has been scanned using continuous XRF analysis and co-registered with hyperspectral core scans. Cluster analysis was used to define host rock types and ore zones based on these data. Sub-sampling strategies were developed, and samples were selected for thin section preparation. Thin sections were inspected using quantitative automated mineralogy to unravel host rock, alteration, and ore zone mineralogy. Results will be integrated with the geology to aid forward and inverse modeling of the geophysical data.

An Experimental Study of Tellurium Solubility in Water Vapor Between 150 - 300 °C: Implications for Ore Formation

Jonathan R. Adams[1], Nicole Hurtig[1], Alexander P. Gysi[1, 2]
1. New Mexico Institute of Mining and Technology, Socorro, NM, USA, 2. New Mexico Bureau of Geology and Mineral Resources, Socorro, NM, USA

Tellurium is becoming of increasing importance in green energy technologies in the production of photovoltaic cells, semiconductors, metal-alloys and is considered a critical mineral. Tellurium has extremely low crustal abundances and is rarely enriched to economic levels. Tellurium is a common constituent in Cu-porphyry and epithermal systems, occurring as telluride minerals and a trace component in sulfides and sulfosalts, that can be considered a byproduct of mining precious and base metals. Prediction of the role of tellurium mobility and speciation in vapors depends on the availability of thermodynamic properties of gaseous Te species at temperatures and pressures prevailing in magmatic-hydrothermal systems. Several gaseous tellurium species are known including Te_g, $Te_{2,g}$, H_2Te_g, $TeCl_{x,g}$, $TeF_{x,g}$, TeO_g, $TeO_{2,g}$, $TeO_2(H_2O)$, and $TeO_2(H_2O)_2$. For most species thermodynamic data are available to predict volatility, however magmatic-hydrothermal vapors are dominantly H_2O, which acts as a solvent promoting metal solubility with increasing temperature and vapor density. Previous experimentally derived thermodynamic properties for hydrated $TeO_{2,g}$ indicate log fugacities ranging from -5.73 to -3.66 between 450 and 700 °C. Previous numerical models of natural systems implicate H_2Te_g will be the dominate gaseous Te species in magmatic-hydrothermal systems. Here we present tellurium solubility experiments determined in pure water vapor between 150 and 300 °C using batch-type Ti-autoclaves. TeO_2 and native tellurium with oxide mineral buffers (MoO_2-MoO_2 and WO_2-WO_3) will be used to test the role of oxidation states of different gaseous Te species on the overall tellurium solubility. Experiments will be conducted in a high-temperature muffle furnace for ~ 15 days, after which quenched condensates and reacted solids will be extracted and analyzed using ICP-MS for determination of Te concentrations and SEM/EPMA/XRD for solid characterization. Thermodynamic properties will be extracted from these experiments and implemented into the GEMS code package for simulating the factors controlling tellurium mobility in ore-forming systems.

Solubility of Cu, Ag and Au in Magmatic Sulfur-bearing Fluids as a Function of Oxygen Fugacity

Alice Alex[1], Zoltan Zajacz[2]
1. Life Cycle Geo, Toronto, ON, Canada, 2. University of Geneva, Geneva, Switzerland

Magma-derived fluids containing chlorine and sulfur are critical for the transport of ore metals to porphyry ore-forming environments. The effect of redox conditions on ore metal speciation and solubilities in sulfur-bearing fluids required for efficient exploration models have not yet been experimentally constrained. We performed experiments to determine the effect of oxygen fugacity (fO_2) on the solubility of Cu, Ag and Au in high-temperature, low- density, low-salinity fluids and hypersaline brine. The experiments were conducted at T=900°C, P=2000 bar varying fO_2 in 7 steps between NNO - 0.5 to NNO + 2.5. The apparent solubility of Cu and Ag at the imposed metal activities increases by a factor of 7 in the H_2O-NaCl-KCl-S low-salinity fluid with fO_2 increasing from NNO - 0.5 to NNO + 2.5. The addition of 0.198 m HCl increases the overall solubility of Cu by a factor of 1.2 – 2.4. The apparent solubility of Au decreases by a factor of 9 as fO_2 changes from NNO - 0.5 to NNO + 2.5. The relationship between the logarithms of the apparent Cu and Ag solubilities and fO_2 is linear and the slope of the fitted line corresponds to 1+ oxidation state of these metals indicating that they are dominantly complexed by ligands that are S-free (e.g., chloride).Considering previous studies on silicate melts and our experimental data for volatiles, Cu and Ag likely have constant fluid/melt partition coefficients in the typical fO_2 range of arc magmatism. Our experiments also show that Cu extraction would be less efficient in S-rich systems at intermediate fO_2. But, the direct exsolution of hypersaline fluids from silicate melts at oxidizing conditions may be a particularly efficient way of Cu sequestration from crystallizing upper crustal magma bodies provided that these brines can be physically mobilized and transferred to the hydrothermal system.

Mineralogy and Trace Elements Chemistry of Quartz Vein from Tsunheg Tungsten Deposit in Western Mongolia

Baatar Amitan[1], Erdenebayar Jamsran[1], Ogata Takeyuki[4], Byambajav Amarjargal[1]
1. MUST, Ulaanbaatar, Mongolia, 2. 鉱物資源科学ラボ, Akita, Japan, 3. Mongolian University of Science and Technology, Ulaanbaatar, Mongolia, 4. MiReS Lab, Akita, Japan

The Tsunkheg deposit is tectonically located in the Mongol Altai and Khovd deep fault zones and it's characterized by the development of numerous faults to the northwest and northeast. The geology of this area is occupied by Upper Ordovician-Lower Silurian Bakhlag formation with sandstone, mudstone, and tuff conglomerate, Shargovi Silurian sedimentary rocks cut by upper Silurian diorite, and small bodies of Carboniferous diabase. The Tsunkheg deposit consists of 6 large quartz–wolframite veins, 0.15-0.45 m wide, 400-600 m long, and one stock body with a diameter of 80 m in the central part. The quartz-tungsten veins contain scheelite, sulfide minerals: chalcopyrite, bornite, arsenopyrite, molybdenite tetrahedrite; carbonate minerals: siderite, malachite, azurite, fluorite, beryllium. Cathodoluminescent (CL) texture and trace element signature of quartz from the Thunkheg deposit has been investigated to discuss their genetic significance. Results show that the pre-ore quartz (QI) displays a bright and homogeneous luminescence texture, (QII) generally shows a bright and oscillatory luminescent texture and, the syn-ore quartz (QIII) generally has a dark and homogeneous (or slightly mottled) luminescent texture that may result from annealing of original CL textures; the post-ore quartz. EPMA analyses suggest Ti, Al, K, Fe, and Ca are the most abundant elements in various generations of quartz and mainly occur as solid solutions within the crystal lattice, in which Al can be incorporated into quartz by substituting Si in the crystal lattice, with additional cations to keep the charge balance. The low Ti concentration and oscillatory euhedral growth zones of CL textures suggest that syn-ore quartz (QII) may have precipitated from a medium-temperature hydrothermal fluid. This study highlights that CL textures and trace element chemistry of quartz related to mineralization can be used to fingerprint the physical and chemical history of ore deposit formation.

Airborne Magnetic and Radiometric Data Provide Insights to Alkaline Intrusive Complexes and Associated REE and Thorium Resources in the Wet Mountains, Colorado

Eric D. Anderson[1], V.J.S Grauch[1], Benjamin Magnin[1, 2]
1. US Geological Survey, Denver, CO, USA, 2. Colorado School of Mines, Golden, CO, USA

The USGS Earth Mapping Resources Initiative (EarthMRI) is tasked with improving the Nation's understanding of the geologic framework within which undiscovered critical mineral resources may occur. To help address this goal, in the summer of 2021 an airborne magnetic and radiometric survey was flown over the Wet Mountains region in south-central Colorado, where several Cambrian alkaline complexes are associated with REE and thorium resources. The survey was flown along east-west flight lines spaced 150 m at a nominal terrain clearance of 80 m and covered an area approximately 2,250 km^2.

The resulting aeromagnetic data and maps of equivalent uranium (U), thorium (Th), and potassium (K) exhibit a unique combination of responses over each alkaline complex that relates to their different petrologic characteristics. Pyroxenite and gabbroic rocks at Gem Park produce large-amplitude magnetic highs and low concentrations of K, Th, and U. In contrast, Democrat Creek, which is largely composed of quartz syenite, exhibits a low magnetic response and high concentrations of K, Th, and U. McClure Mountain complex, a mixture of syenites and mafic rocks, shows a complicated magnetic pattern, high K, and low U and Th. The adjacent stratified mafic-ultramafic rocks of Iron Mountain exhibit local magnetic highs and overall low radiometric response. North-northwest trending Th anomalies correlate well with previous mapping of Th-rich veins and dikes and suggest additional structures south of the outcropping alkaline complexes. Farther to the southeast, a magnetic high having similar amplitude and form to that observed over the Gem Park and Iron Mountain intrusions suggests additional gabbro and pyroxenite rocks in the shallow subsurface. Geophysical characterization of the alkaline intrusions and veins help guide the interpretation of their subsurface extents and their possible association with faults, which are also well expressed in the aeromagnetic data as breaks in anomaly patterns.

Translithospheric Tracers of Ni-Cu-Co-PGE Mineralisation: a Geochemical Approach

Daryl E. Blanks[1], David A. Holwell[1], Marco L. Fiorentini[2]
1. University of Leicester, Leicester, United Kingdom, 2. University of Western Australia, Perth, WA, Australia

Sulfide deposits hosted by mafic-ultramafic magmatic rocks are an important source of Ni, Cu, Co and PGEs; metals essential for renewable energy (e.g. wind turbines) and electric vehicles batteries. However, discoveries of such deposits are decreasing and thus locating new metal resources is critical for ensuring a security of supply for green energy technologies.

Magmatic systems play a crucial role in enriching the crust with metals that reside primarily within the Earth's mantle e.g. Ni, Cu, Co, Pt, Pd, Te and Se. The nature of the mantle source, degree of partial melting, and tectonic setting is widely thought to control the initial metal and S concentrations of these ascending melts during the primary stages in the generation of Ni-Cu-PGE mineralisation. The behavior of chalcophile elements and partitioning effects between sulfide liquid, silicate melts, and hydrothermal fluids at varying depths in the lithosphere make them useful tracers for a variety of mantle and crustal processes.

Recent advances in the understanding of magmatic mineral systems have highlighted continuums between different sulfide deposit styles from the magmatic Ni-Cu systems to porphyry Cu-Au and hydrothermal Cu-Ni-Au through the lithosphere. We show how distinct variations in whole rock metal ratios (e.g. Ni/Cu/Pd/Te/Se), from a number of different mineral systems can be used as tracers of different lithospheric processes from initial metal enrichments during partial melting of the mantle, through sulfide fractionation on ascent and hydrothermal mobilization in the upper lithosphere. This appreciation of the divergent mantle-crustal metallogenic signatures can broaden our current understanding on the mineralising stages during magmatic sulfide genesis and help to vector towards ore deposits in distal expressions of mineralised magmatic systems.

Lead Isotope and Fluid Inclusion Investigations of the Hicks Dome Critical Mineral Resource, Illinois-Kentucky Fluorspar district, USA

Mitchell M. Bennett[1], Albert H. Hofstra[1], Mike Pribil[1], Allen K. Andersen[2], Wayne Premo[3], Anne E. McCafferty[1], Anton Maria[4], Laurence M. Nuelle[5]
1. US Geological Survey, Denver, CO, USA, 2. US Geological Survey, Spokane, WA, USA, 3. US Geological Survey (Emeritus), Denver, CO, USA, 4. University of Southern Indiana, Evansville, IN, USA, 5. Hicks Dome LLC, Hardin County, IL, USA

Deep drilling at Hicks Dome has defined an inferred critical mineral resource of 65.8 Mt with average grades of 15.8% fluorite, 4.8% barite, 1.3% TiO_2, 0.3% REY oxide, 0.2% Nb_2O_5, and 0.16% BeO using a fluorite cut-off grade of 10%. Fine-grained purple fluorite mineralization occurs in solution collapse breccias with a rock flour-matrix. Breccias are hosted in Ordovician to Silurian platform carbonate rocks at depths of 700-1000 m. The resource is associated with 271 Ma lamprophyre dikes and diatremes, an upper crustal magnetic anomaly, and a Th radiometric anomaly at the surface. It is hypothesized the origin for the dome and the HREE enrichment is the product of fluid exsolved from a concealed carbonatite intrusion.

SEM-EDS raw x-ray intensity maps and laser Raman spectroscopy were utilized to identify mineral phases in high grade ore. X-ray intensity maps were grouped with Aztec algorithms to form phase maps that identify fluorite, barite, Pb-Zn-Fe-sulfides, Ca-Mg±Sr±Ba carbonates, apatite, Th-REE phosphates, and Nb-Ti oxides. In-situ Pb isotope analyses of ore and gangue minerals by LA-MC-ICP-MS is proceeding to establish an isochron and to constrain the source(s) of Pb in the system.

Quartz and fluorite contain distinct secondary fluid inclusion populations. Two-phase liquid-rich inclusions are irregular to equant in shape, with a small to medium bubble. Three-phase liquid-vapor-solid inclusions are morphologically similar to the liquid-rich inclusions. Sparse vapor-rich inclusions occur with liquid-vapor inclusions in fluorite. Three-phase water-CO_2 inclusions are restricted to quartz, contain varying ratios of vapor- and liquid-CO_2 and liquid water; some assemblages contain a solid daughter mineral. Pyrite mineral inclusions are restricted to late-stage fluorite, while primary liquid-rich inclusions are present in calcite and late-stage fluorite. Microthermometric and laser Raman characterization of these fluid inclusion populations is underway to constrain the temperature, salinity, and gas compositions of hydrothermal fluids in the system.

Exploration and Mining Industry Feedback on the USGS Earth Mapping Resources Initiative (EarthMRI)

Andrew R. Giebel
Colorado Geological Survey, Golden, CO, USA

The Colorado Geological Survey, in partnership with the United States Geological Survey, is conducting a national survey of mining and exploration professionals regarding the responsible development of our Nation's critical and strategic mineral resources. The United States remains heavily dependent on imports of mineral commodities vital to the Nation's economic and national security interests.

Our aim is to improve our understanding of the industry's current mindset, data needs, and general awareness of the EarthMRI program and its published datasets. USGS and state geological surveys will use the information provided to help guide public investment into mapping, sampling, and other data collection programs that are most valued by the exploration community, and which encourage private investment in the targeting of critical and strategic mineral resources domestically. Additionally, we hope to connect and open a dialogue with individuals and firms that may be further interested in partnering with the USGS or local state geological surveys to identify specific areas where public investment might further encourage private exploration activity.

Defining Volcanic Stratigraphy and Syn-volcanic Intrusions at the Lynne Zn-Pb-Cu Deposit, Wisconsin, USA

Lillian N. Glodowski, Robert W. Lodge, Shelby R. Short
University of Wisconsin-Eau Claire, Eau Claire, WI, USA

The Lynne Zn-Pb-Cu deposit was discovered in 1990 by Noranda Exploration and is one of many volcanogenic massive sulfide (VMS) deposits located in northern Wisconsin within the Paleoproterozoic Penokean volcanic belt. The environment for VMS deposition has previously been interpreted as continental back-arc volcanism due to the accretion of the Pembine-Wausau Terrane onto the Archean Superior Craton during the Penokean Orogeny (1.8-1.9 Ga). However, the deposit-scale geodynamic setting and its influence on the metallogeny of VMS deposits are poorly constrained. This project constrains the volcanic and tectonic setting at the Lynne Zn-Pb-Cu deposit in Oneida County, Wisconsin, and contributes to our understanding of the variations of VMS-forming environments within the Penokean volcanic belt. Samples of least-altered volcanic and intrusive rocks from the deposit were obtained from drill core and were prepared for geochemical and petrographic analyses. The volcanic units at the Lynne deposit were previously divided by their stratigraphic position relative to the ore horizon into upper and lower rhyolite/dacite units. Major and trace element geochemistry and petrographic observations from these volcanic units reveal no petrochemical difference between the upper and lower strata, indicating that VMS formation is not associated with transitioning geodynamic settings. Intersecting the Lynne deposit are fine grain felsic dikes and a coarse grain granodiorite pluton at the stratigraphic footwall of the deposit. These intrusive rocks at the Lynne deposit are geochemically indistinguishable from the volcanic host rocks which suggests they are a part of the same magmatic system. Additionally, the geochemical similarities to the felsic volcanic rocks and lack of contact metamorphic aureole suggest the granodiorite is the syn-volcanic intrusion that may have helped drive hydrothermal circulation prior to its final emplacement.

Onganja, Namibia: to Be or Not to Be an IOCG?

Matthew P. Hales[1], Judith Kinnaird[1, 2], Paul Nex[1, 3]
1. University of the Witwatersrand, Johannesburg, South Africa, 2. NRF/DSI Centre of Excellence (CIMERA), Johannesburg, South Africa, 3. Economic Geology Research Institute (EGRI), Johannesburg, South Africa

The Onganja Mining District, ~60 km northeast of Windhoek, Namibia is a world-renowned gem cuprite deposit. Cuprite, however, has only contributed to a small portion to the life-of-mine as exploitation dates back several centuries.

Formal mining began in 1904 and targeted several high-grade (2-6% Cu), low tonnage (<10000t) orebodies. These orebodies comprised subvertical breccias, consisting of chalcocite and chalcopyrite, associated with a N-S striking, sheeted-vein complex and subhorizontal, stratabound lenses, consisting of chalcopyrite, within the Kuiseb Schist Formation of the Damara Orogeny Southern Zone. Locally, gold and molybdenite are present but do not show a strong correlation to copper mineralization. Gangue minerals include quartz, albite, biotite, chlorite, calcite, magnetite and hematite with minor rutile, apatite and scapolite being locally present. These assemblages have led to the hypothesis that the district represents an IOCG system.

Alteration assemblages mimic conventional IOCG assemblages with albite and biotite representing Na^+ and K^+ alteration, respectively, but neither are regionally extensive. Rather, these minerals form in distinct horizons or as subhedral crystals within unmineralized breccias. Likewise, magnetite and hematite are locally prevalent, however, the formation of these minerals appears to have been controlled by the country-rock composition and degree of albitization; the same is true for chlorite with the added complexity that chlorite is also a product of retrograde metamorphism within the country schists. Scapolite is regionally extensive throughout the Southern Zone but regionally is more meionitic as compared to that at Onganja.

Furthermore, there are no known intrusions near the district, and the high-P, low-T conditions of the Southern Zone are the opposite of the conditions favourable to metamorphic IOCG systems.

Given these complexities, the Onganja mining district may represent an IOCG, however, the interplay between orogenic processes and ore-forming processes means that alternatives, such as remobilised VMS, cannot yet be discounted.

Towards a Genetic Model for the Cu-(Te-Au-Ni-PGE) Sulfide Deposits of the Curaçá Valley, Brazil

David A. Holwell[1], Daryl Blanks[1], John Thompson[2], Ingrid Mota[3], Filipe Porto[3]
1. University of Leicester, Leicester, United Kingdom, 2. PetraScience Consultants, Vancouver, BC, Canada, 3. Ero Copper, Vancouver, BC, Canada

The Curaçá Valley hosts a suite of Cu-rich sulfide deposits hosted by ultramafic/mafic intrusions, within the Itabuna-Salvador-Curaçá Orogenic Belt, Brazil. The origin of the deposits is contentious, with Cu-rich fractionated magmatic sulfide, hydrothermal/metamorphosed magmatic sulfide, alkalic magmatic sulfide and iron-oxide-copper-gold (IOCG) origins all suggested. Key aspects that set the Curaçá Valley deposits aside from other ultramafic-mafic hosted sulfide deposits are: the very high Cu/Ni ratios of most of the deposits; the extreme Te enrichment; the intimate association of the sulfides with volatile phases, in particular, phlogopite; and the depth of preservation (deep, lower crust). We rule out an IOCG origin based on geochemical characteristics of the ores, in particular, the low Cu/Te ratios and high PGE contents compared with IOCGs, and prefer a magmatic origin. The depth of preservation opens up the possibility the sulfides may represent magmatic Ni-Cu-PGE sulfides generated from ultramafic/mafic melts that have undergone lengthy and extensive fractionation in a 'Goldilocks zone' or 'temperature window'; producing extreme Cu-Te-Au-rich end members, and implying Ni-PGE-rich end members elsewhere in the system. Alternatively, or in addition to, the presence of abundant volatile phases (phlogopite, apatite, carbonate) along with the high Te content and chalcophile metal concentrations and ratios, is also entirely consistent with an alkalic magmatic signature, where the parental magma was sourced from low degree partial melts of an enriched mantle source. Here we utilise a combination of mineralogical, isotopic and geochemical techniques to interrogate these models and, by comparing to other analogous deposits worldwide, aim to define a sub-class of magmatic sulfide deposits that may be characteristic of lower crustal magmatic systems

Geochemistry and Genesis of the Manganese Deposits in Mankwadzi Southern Kibi-Wenniba Belt, Ghana

Kwabina Ibrahim, Frank K. Nyame, Johnson Manu
University of Ghana, Accra, Ghana

ABSTRACT

Major and trace elements distribution in samples from the manganese occurrence in the Mankwadzi area on the southern Kibi-Winneba belt of ghana were investigated and the results obtained were statistically analyzed. The results show that the samples are enriched in Cu, Pb, Zn, and Ba and depleted in Co, Ni, Be, Mo, Sr, and Sn. MnO correlates negatively with Fe_2O_3 and the componentsK_2O, Al_2O_3, MgO, Cu, Co, Zn, Sr, and Sn correlated positively with Mn. SiO2, Be, and Mo show a positive correlation with Fe and correlated negatively with Mn. Factor analysis extracted a five-factor model, that could explain ca 80% of the total variance in the data. Fi is interpreted as a separation factor between the Fe and Mn mineral. Fe2 and Fe3 reflect the association of some trace elements (Cu, Co, Zn, Ni, Pb, Sn, and Sr) with Mn minerals. F4 is a dolomite factor (Cao Mg O) referring to the unreplaced remnants of the original country rocks. F5 reflects the dispersion of Ba in both ore minerals and country rocks. A comparison of the chemical composition of rocks with that of ferromanganese deposits occurring in various present-day environments suggests that hydrothermal activity is mainly responsible for their formation.

Mineralogical Sequestration of Tellurium at the Perseverance Volcanogenic Massive Sulfide Deposit, Quebec, Canada

Filip Kasprowicz[1], Katharina Pfaff[1], Thomas Monecke[1], David Diekrup[2], Mark Hannington[2]
1. Center for Mineral Resources Science, Department of Geology and Geological Engineering, Colorado School of Mines, Golden, CO, USA, 2. Department of Earth and Environmental Sciences, University of Ottawa, Ottawa, ON, Canada

Tellurium is a rare element that is essential for the production of photovoltaic cells. At present, most Te is produced as a by-product of anode slimes from electrorefining of copper ores. Volcanogenic massive sulfide (VMS) deposits represent one of the few primary sources of Te. Kankberg in the Paleoproterozoic Skellefte district in Sweden and the Silurian to Devonian VMS deposits of the Urals in Russia collectively account for ~17% of the global Te production of ~450 tonnes Te. While deposits of this type are significant producers of Te, there is a knowledge gap pertaining to the origin and enrichment processes of Te in VMS deposits and the mineralogical sequestration of this element in massive sulfide ores. To close this knowledge gap, this study focuses on the ore mineralogy of the Perseverance VMS deposit in Quebec, Canada, which was known for its elevated Te content. The deposit was mined between 2008 and 2013 with pre-mining resources of 5.1 million tonnes of ore grading at 15.8% Zn, 1.24% Cu, 29.4 g/t Ag, and 0.4 g/t Au.

Polished thin sections of samples collected from different ore zones were investigated by quantitative scanning electron microscopy to locate Te-bearing minerals and to determine their relative abundances. Minerals identified include altaite, hessite, montbrayite, frohbergite, and melonite. The microtextural characteristics of these phases were subsequently studied by scanning electron microscopy and reflected light microscopy. Many of these phases are located along the grain boundaries of the main ore minerals or form small inclusions within them. The textual evidence suggests that the Te-bearing phases formed primarily during metamorphic recrystallization and remobilization as impurities originally contained in the major ore minerals were expelled from the crystal structure during recrystallization to form new accessory minerals. These changes in mineralogical sequestration have important implications to mineral processing.

Nickel Mineralogy of the Historic Callenberg Ni-Laterite District, Saxony, Germany

Alkiviadis Kontonikas-Charos[1], Max Frenzel[1], Mathias Burisch[1, 2], Jens Gutzmer[1]
1. Helmholtz Institute Freiberg for Resource Technology, Freiberg, Germany, 2. TU Bergakademie Freiberg, Institute of Mineralogy, Freiberg, Germany

With the ever-growing demand for technology-critical resources (TCR), there is renewed interest in potentially underexploited historic occurrences of these metals. The Callenberg laterite district, Saxony, comprises the Callenberg North and Callenberg South sub-deposits, and was a vital source of Ni for East German industry during the late 20th century. Approximately 9.15 Mt of ore was extracted between 1960-1990, with average grades of 0.8-1% Ni. At present, an estimated resource with 34,500 t of contained Ni lies nearby at Kieferberg. Following mine closure, few studies on the Callenberg district were published (including the first description of gem quality crocoite specimens), however many of these are inaccessible and limited to German academic archives. Here we summarize the geology of the Callenberg laterites, and provide our own findings on Ni mineralogy in samples from the TU Bergakademie Freiberg Geoscience Collection.

The Callenberg district occurs within a 7 km NW-SE striking serpentinite complex in the SW Granulite Mountains, Saxony. Initial uplift of Variscan granulites during the Mesozoic led to significant weathering and lateritization throughout the Late Tertiary. Unlike most laterite deposits, Callenberg is a rare example of preserved ancient paleosols, found well outside the present-day tropics. Covered by up to ~20 m of overburden, the laterite ores vary in thickness from 1-20 m. They are classified as hydrous silicate laterites, and comprise serpentinized peridotite bedrock overlain by lower saprolite (hard, silicate-rich), upper saprolite (earthy), and goethite + hematite-rich limonite layers. Lower saprolite is Fe-poor (<10% Fe) and predominantly consists of Mg-chlorite, smectite, serpentine, magnetite, maghemite, chromite, and Mn-oxides. In contrast, the upper saprolite layer is Fe-rich (>10% Fe) due to abundant goethite, maghemite and hematite occurring with various phyllosilicates (chlorite, smectite and serpentine). The main hosts of Ni in saprolite are garnierite, which forms veins with quartz ± chalcedony, and serpentine minerals (lizardite-nepouite).

Using Ore Petrography and Geochemical Mass Balance to Constrain the Hydrothermal Environment at the Paleoproterozoic Flambeau Cu-Zn-Au Deposit, Wisconsin, USA

Robert W. Lodge, Tara C. Lemke, Kaelyn E. Blotz
University of Wisconsin-Eau Claire, Eau Claire, WI, USA

Volcanogenic massive sulfide (VMS) deposits can form in a variety of tectonic environments that ultimately influence the composition of ores and style of hydrothermal alteration. Traditional models for VMS formation factor in the influence of host rock and hydrothermal fluid composition on ore types and alteration products. These subtle differences in ore minerals and hydrothermal products are important for understanding the genesis of VMS deposits in various environments and has served as a first-order exploration model. The Paleoproterozoic Flambeau Cu-Zn-Au deposit in northern Wisconsin produced 1.9 Mt of supergene-enriched ore but the deposit geology and ore genesis has not been fully characterized because of poor exposure and a lack of mineral investment. However, the deposit has unique alteration assemblages (biotite-andalusite-chlorite schists) and is one of the most Cu- and Au-enriched deposits in the Penokean Orogen. Therefore, improving our understanding of the hydrothermal features of this deposit will help improve our knowledge of the fluid conditions that from Cu- and Au-rich VMS deposits.

This study sampled representative metamorphosed hydrothermally-altered rocks and ore zones of the Flambeau deposit from drill core stored at the Wisconsin Geologic and Natural History Survey. Whole rock lithogeochemistry from altered rocks reveal characteristic major element mobility found in most VMS deposits. However, mass balance calculations reveal an enrichment of potassium and silica that are characteristic of argillic alteration and have little evidence for a significant component of Mg-metasomatism that is common in many other VMS deposits in similar tectonic environments. Additionally, SEM-EDS of polished ore sections have identified various tellurides (hessite, altaite, tsumoite, bismuth), electrum, arsenopyrite, acanthite, bismuthinite, cassiterite, monazite, and an unnamed tungsten mineral. This suite of trace minerals suggests a magmatic fluid component to the hydrothermal system. Based on these observations, the Flambeau deposit is possibly an example of a metamorphosed hybrid VMS-epithermal system.

Structural and Kinematic Analysis of Alkaline Dikes and Mineralized Veins in the Wet Mountains, Colorado: Connecting Rare Earth Element Mineralization to Cambrian Rifting

Benjamin P. Magnin[1, 2], Yvette D. Kuiper[1], Eric D. Anderson[2]
1. Colorado School of Mines, Golden, CO, USA, 2. United States Geological Survey, Denver, CO, USA

We investigated the structural controls on, and kinematics during, emplacement of Cambrian-Ordovician, failed-rift-related alkaline intrusive rocks and associated thorium and rare earth element (REE) mineralization exposed in Precambrian metamorphic rocks of the Wet Mountains, CO. Results were compared to other nearby Cambrian-Ordovician aulacogens to determine if they are related. Alkaline intrusive rock in the Wet Mountains comprise the McClure Mountain, Gem Park, and Democrat Creek complexes, which are sequentially cross-cut by lamprophyre, syenite, and carbonatite dikes, and mineralized quartz-barite-thorite veins. Syenite dikes contain hydrothermal alteration hypothesized to be temporally associated with the latest stage mineralized veins. REE mineralization occurs mainly within carbonatite dikes, quartz-barite-thorite veins, and hydrothermally altered red syenite dikes. Structural data show that all types of dikes/veins strike dominantly northwest and dip steeply. Outcrop observations of structural fabrics support four principal styles of emplacement: (1) fault-controlled emplacement indicated by adjacent brecciated and gouged wall rock, (2) tensile, fracture-controlled emplacement based on no associated wall rock deformation, (3) emplacement along reopened (Precambrian?) pegmatite dikes, and (4) local emplacement along Precambrian metamorphic foliation. These observations are consistent with emplacement along northwest-trending tensile fractures and high-angle faults formed in a northeast-southwest-directed extensional regime. Kinematic analysis of mineralized high angle faults and tensional fractures indicate that extension was horizontal and 045°/225°-trending. The northwest trend and extensional nature of the mineralized dikes and veins suggest that the Cambrian-Ordovician alkaline rocks of the Wet Mountains formed in a rift setting and may be a continuation of the northwest-trending, Cambrian southern Oklahoma aulacogen, the only nearby Cambrian along-strike rift system with the same orientation as the alkaline rocks in the Wet Mountains. Considering the potential relationship between the Wet Mountains and the southern Oklahoma aulacogen, future REE exploration should focus along the entire extensional system.

The Lanthanide Tetrad Effect as an Exploration Tool for Granite-related Rare Metal Ore Systems: Examples from Iberian Variscides

Ivo Martins[1, 2], António Mateus[1, 2], Isabel Ribeiro da Costa[1], Miguel Gaspar[1], Ícaro Dias da Silva[1, 2]
1. Departamento de Geologia, Faculdade de Ciências da Universidade de Lisboa, Lisboa, Portugal, 2. Instituto Dom Luiz, Faculdade de Ciências da Universidade de Lisboa, Lisboa, Portugal

The current technological evolution and energy transition policies amplify the dependence on a large number of metals, many of them with low recycling rates. This generates additional pressure on mineral exploration endeavors to search new primary resources for these strategic metals, which are often associated with highly-evolved muscovite-bearing peraluminous granites. Granite differentiation and related magmatic-hydrothermal ore-forming processes can be traced by elemental content ratios such as K/Rb, Sr/Eu, Y/Ho, Rb/Sr, Nb/Ta and Zr/Hf. The lanthanide tetrad effect ($TE_{1,3}$) is also a useful whole-rock geochemical fingerprint of granite differentiation. Its application as an exploration vector for granite-related mineralization in the Central-Iberian Zone (CIZ) is assessed in this work by examining the $TE_{1,3}$ variations along with those elemental ratios, and with the concentration of Sn, W, Nb, Ta, Li and fluxing elements. The U-Pb zircon dating and multi-elemental whole-rock geochemical characterization of the main plutons and late aplite-pegmatite dykes exposed across the Segura-Panasqueira Sn-W-Li belt (in CIZ) show that Cambrian-Ordovician and Carboniferous-Permian granite suites: (i) display different degrees of differentiation and metal-enrichment; and (ii) their compositional features compare well with data published for similar rocks from other Variscan segments. Increasing $TE_{1,3}$ (up to 1.5) co-varies with magmatic differentiation and metal-enrichment, and the Carboniferous-Permian granites are more differentiated, and metal specialized. The Argemela Li-Sn-bearing Rare Metal Granite (RMG) and the Segura aplite-pegmatite dykes deviate from this geochemical trend, displaying $TE_{1,3} < 1.1$, but also high P_2O_5 contents. The results obtained suggest that mineralized rocks related to Peraluminous-High-Phosphorous Li-Sn(-Nb-Ta) granite systems are typified by $TE_{1,3} < 1.1$, whereas those associated with Peraluminous-Low-Phosphorous Sn-Ta-Nb granite systems display $TE_{1,3}$ well above 1.1, reaching values as high as 1.7-2.1 in the case of the Penouta RMG (NW Spain).

This work was supported by FCT(PD/BD/142783/2018) and FCT/UIDB/50019/2020 – IDL, being also a contribution of MOSTMEG project (ERA-MIN/0002/2019), //mostmeg.rd.ciencias.ulisboa.pt/.

Lithogeochemical Vectors and Mineral Paragenesis of Hydrothermal REE-Bearing Fluorite Veins and Breccias in the Gallinas Mountains, New Mexico

Evan J. Owen[1], Alexander P. Gysi[1, 2], Virginia T. McLemore[2], Nicole Hurtig[1], Katharina Pfaff[3]
1. Dept. of Earth and Environmental Science, New Mexico Tech, Socorro, NM, USA, 2. New Mexico Bureau of Geology and Mineral Resources, New Mexico Tech, Socorro, NM, USA, 3. Dept. of Geology and Geological Engineering, Colorado School of Mines, Golden, CO, USA

The Gallinas Mountains district in New Mexico showcase hydrothermal REE-bearing fluorite veins and breccias hosted in Permian sedimentary rocks and formed during the emplacement of trachyte/syenite sills, dikes and breccias ~30 Ma. Minor production of fluorite, REE, base, and precious metals has been recorded in the early 1900s. Gallinas Mountains is a prime location to study hydrothermal REE mobilization in an alkaline system because of the well exposed geology. REE are found in bastnäsite-(Ce), which is also the primary ore mineral of several important carbonatite REE deposits such as Mountain Pass in California and Bayan Obo in China. The aim of this study is to determine whether REE enrichment in the district can be vectored using fluorite and whole rock geochemistry.

Three REE-bearing fluorite vein types have been characterized using optical microscopy, BSE imaging, SEM-based automated mineralogy, and CL microscopy. These are: (i) barite-fluorite, (ii) bastnäsite-fluorite, and (iii) calcite-fluorite veins. Three distinct fluorite generations (Fl I-III) with unique CL signatures were distinguished in these veins. Of these, Fl II, found in type (ii) veins, appears to be most significant to REE deposition as it forms fine intergrowths with bastnäsite-(Ce) and these veins overprint Fluorite I in type (i) veins. Preliminary LA-ICP-MS analysis on fluorite indicates distinct REE chondrite-normalized profiles for each fluorite type. Fluorite I exhibits a LREE-enriched profile, Fluorite II a flat LREE profile depleted in HREE, and Fluorite III a LREE-depleted and HREE-enriched profile. Whole rock F and REE of the fluorite veins and breccias display a positive correlation with Ba, which indicates an increase in REE mineralization associated with the barite-fluorite veins. These new results indicate that F-metasomatism plays a key role in the hydrothermal mobilization and deposition of REE, which needs to be further investigated to develop geochemical vectors in this type of REE mineral deposit.

Presence of In, Ga and Ge in the Epithermal High Sulfidation Deposit of La Mejicana Mine, Famamtina District, NW Argentina

Morena L. Pagola[1, 2], Sebastián M. Jovic[1, 2, 3], Diego M. Guido[1, 2]
1. Instituto de Recursos Minerales, Facultad de Ciencias Naturales y Museo, Universidad Nacional de La Plata, La Plata, Argentina, 2. Consejo Nacional de Investigaciones Científicas y Técnicas, La Plata, Argentina, 3. Departamento de Geología. Universidad de Buenos Aires, Buenos Aires, Argentina

The La Mejicana high sulfidation Cu-Au epithermal veins are located in the Famatina mining district, northwestern Argentina.The district contains high sulfidation Cu-Au-(As-Sb-Te) epithermal veins and small Cu-Mo-Au-(As-Sb-Te) mineralized porphyries of the early Pliocene age (Losada-Calderón et al.1994). The district has been known since the mid 19th century and was one of the most important copper, gold and silver producers until 1926. The mining activity has yielded more than 1 Mt of ore, average 3 percent Cu, 80 g/t Ag, and 11 g/t Au from more than 50 veins. The ore is composed of famatinite, enargite, pyrite, quartz, and minor tennantite, tetrahedrite, sphalerite, gold, tellurides, covellite and chalcopyrite.

Sahlström et al. (2017) analyzed samples of enargite (ICP-MS-LS) and obtained average Ge concentrations of 717 ppm, and low anomalies for In and Ga. In this abstract we present new ICP analysis of ore samples that shows indium anomalies up to 419 ppm and gallium and anomalies up to 109 ppm. High correlation between In-Ga and Zn is present, this could be related with the presence of sphalerite in the ore. This preliminary data indicates that the enargite could be the main hosting mineral for Ge while In and Ga could be present in the sphalerites. Actually, more detail study is ongoing in order to determine the distribution and presence of these critical elements.

Losada-Calderón, A.J., McBride, S.L., and Bloom, M.S., 1994. The geology and 40Ar-39Ar geochronology of magmatic activity and related mineralization in the Nevados del Famatina mining district, La Rioja Province, Argentina: Journal of South American Earth Sciences, v. 7, p. 9–24.

Sahlström, F., Arribas, A., Dirks, P., Corral, I., & Chang, Z., 2017. Mineralogical distribution of germanium, gallium and indium at the Mt Carlton high-sulfidation epithermal deposit, NE Australia, and comparison with similar deposits worldwide. Minerals, 7(11), 213.

Characterization of the Footprint of Hydrothermal Ore Forming Processes Through Trace Metal Signatures in Fluorite in the Cooke's Peak Pb-Zn-Ag-F District, New Mexico

Cody D. Schwenk[1], Nicole C. Hurtig[1], Alexander P. Gysi[1, 2], Virgil Lueth[2]
1. New Mexico Institute of Mining and Technology, Socorro, NM, USA, 2. New Mexico Bureau of Geology and Mineral Resources, Socorro, NM, USA

The Cooke's range in Grant and Luna Counties, New Mexico, is host to multiple mining districts that have produced over 45 Kt fluorite, 3.6 Kt Pb, 2.7 Kt Zn and 1.98 t Ag since its discovery in 1876. In the southern part of the district, Fluorite Ridge and the northern part, Northern Cooke's district veins are dominantly fluorite with minor base and precious metals and their formation has been classified by recent workers as Rio Grande Rift type deposits formed within the last 10 Ma. The Central and Hadley districts have a higher proportion of base and precious metals and commonly show several generations of fluorite, which have been attributed to hydrothermal systems related to the emplacement of the Cooke's Peak Granodiorite, or other Eocene magmatism. Here we present LA-ICP-MS trace element signatures in different fluorite samples in combination with fluid inclusion petrography and microthermometry to provide a comprehensive ore formation model of the Cooke's Peak Pb-Zn-Ag-F district. Fluid inclusions that were identified include two phase liquid-vapor (8-10 and vol%. vapor) with ice melting temperatures of -2.0 to -0.1 °C indicating the presence of low salinity aqueous fluids (3.74-0.2 wt.% NaCl) with homogenization temperatures ranging from 150 to 190 °C. Five fluorite types were characterized using cathodoluminescence and petrographic observations, and their trace metal signatures show REE patterns that are distinct by generation and district. Heavy and light REE fractionation is observed on a regional trend in fluorite from Fluorite Ridge, Northern, and the early generation from the Central district. In the Central district, heavy REE are increasingly depleted from early to late fluorite generations. Lead, zinc copper and silver are higher in the early fluorite in the Central zone and abundant at Fluorite Ridge indicating that those fluorite generations might be associated with magmatic-hydrothermal base and precious metal mineralization.

Magmatic Controls on Platreef Stratigraphy, Northern Limb of the Bushveld Complex, South Africa

Erin S. Thompson[1], David Holwell[1], Andy Lloyd[2]
1. Centre for Sustainable Resource Extraction, University of Leicester, Leicester, United Kingdom, 2. Anglo American South Africa, Johannesburg, South Africa

The Platreef, northern limb of the Bushveld Complex, South Africa, is globally renowned as one of the world's largest resources of platinum-group elements (PGEs), with significant additional base metal (Ni, Cu, Co) mineralisation. These metals are essential to the growth of sustainable and environmentally friendly technologies, particularly within the electric automotive industry, a market which is expected to grow exponentially in the coming decades.

The Platreef has PGE grades comparable with the Merensky Reef of the eastern and western limbs of the Bushveld, typically 3-4 g/t (Pt + Pd + Rh + Au), but can increase to 15 g/t. The mineralisation is also present over a much greater thickness, tens of metres, compared to <1 m in the Merensky Reef, along with substantially more Ni-Cu-Co in the form of magmatic sulfides. Better understanding the stratigraphy of the Platreef, both in terms of along strike and down-dip variations, is vital for understanding the controls on grade formation, developing genetic and exploration models, which will aid the prediction of the locations of economically important horizons.

The Platreef is thought to have formed via the emplacement of discrete magmatic units or sills, each with their own metal budgets, thicknesses and geochemical characters. Studies have used bulk geochemistry, Ni:Cu and Pd:Pt ratios, mineralogy and platinum group mineral assemblages to differentiate between these packages in the southern and central sectors of the limb. This study uses downhole major and trace element geochemical data to identify and correlate discrete magmatic units at various locations in the central sector of the Platreef. Several units are identified which can be correlated up dip towards the footwall contact to the north east. The control that this primary magmatic stratigraphy exhibits on grade distribution has been examined, with some units carrying considerable grade, compared to negligible grade in others.

Using La-Icp-Ms Trace-element Sulphide Mapping to Delineate Multiphase Co-Cu Mineralization in the Kakanda Deposits, Central African Copperbelt DRC

Helen Twigg[1], Murray W. Hitzman[1], Maurice Brodbeck[2]
1. iCRAG, University College Dublin, Dublin, Ireland, 2. Trinity College Dublin, Dublin, Ireland

The Neoproterozoic Central African Copperbelt located in southern Democratic Republic of Congo (DRC) and the northwestern Zambia contains ~50% of the world's cobalt reserves and significant resources of copper, zinc, nickel, and gold. These energy critical elements are hosted in the Katanga Supergroup carbonate and siliciclastic metasedimentary rocks which were deposited in an epicontinental basin formed during the break-up of Rhodinia. Multiphase hypogene mineralization in the Katangan basin occurred over a time frame of >300 million years linked to basin rift and inversion events which also encompassed salt evacuation and dissolution.

The multiple phases of mineralization have been distinguished with the aid of LA-ICP-MS trace element mapping of sulfide minerals and SEM-EDX mapping of associated alteration phases. Early stage carrollite (Car I) exhibits multiple growth stages with Sb-Mo depleted cores overgrown with elevated Sb-Mo overgrowths. Car I has multiple morphologies; it can be inclusion rich or poor and can form anhedral masses or euhedral crystals. Copper sulphide minerals: chalcocite, bornite and chalcopyrite do not exhibit growth zones. They intergrow with and overgrow Car I. Copper sulphide minerals which overgrow fractured Car I are often associated with dolomite-quartz veins and are intergrown with phyllosilicates, talc in carbonate rocks and chlorite or muscovite in siliciclastic host rocks. Locally, a second phase of carrollite (Car II) with a vuggy texture overgrows the previous sulfide minerals. Car II is enriched in Ni-Mo, has an elevated Co:Cu ratio, and is depleted in Sb and Bi relative to Car I. Vugs within Car II are variably filled with sphalerite. Pyrite outboard of the Cu-Co zone in the deposit displays multiple growth zones defined by elevated concentrations of As, Bi, Mo, Ni, Sb and Se. Rhenium values are significantly higher in Car II compared to earlier sulfide phases which suggests potential issues with mixed phase Re-Os dating.

Zircon and Monazite U-Pb Geochronology of Aquamarine-bearing Yamrang Pegmatite from Eastern Nepal Himalaya

Sushmita Bhandari[1, 2, 3], Kezhang Qin[1, 2], Qifeng Zhou[4]
1. Key Laboratory of Mineral Resources, Institute of Geology and Geophysics, Chinese Academy of Sciences, Beijing, China, 2. University of Chinese Academy of Sciences, Beijing, China, 3. Department of Mines and Geology, Kathmandu, Nepal, 4. Institute of Mineral Resources Research, China Metallurgical Geology Bureau, Beijing, China

The Yamrang Pegmatite from the Ikhabu Pegmatite Field of Eastern Nepal Himalaya is Nepal's primary source of aquamarine with five internal zonations. Lack of deformation in the Ikhabu Pegmatite Field indicates that it is younger than Himalayan Miocene metamorphism; however, the age of Yamrang Pegmatite and the Ikhabu Pegmatite Field has not been dated before. This study reports the U-Pb zircon and monazite geochronology from Yamrang Pegmatite and the Ikhabu Pegmatite Field. Zircon grains were separated for U-Pb dating from Yamrang Pegmatite and Wairung Pegmatite of Ikhabu Pegmatite Field. They were anhedral to euhedral with inherited cores and oscillatory rims, with an average size of around 100 μm. The Wairung Pegmatite sample shows a mean age of middle Miocene (26 zircon spots, 13.6 Ma, MSWD = 0.8). However, two samples from Yamrang Pegmatite about 15m apart show a large age discrepancy, which contradicts geological evidence that Yamrang Pegmatite is younger than Himalaya's Miocene deformation. The Yamrang Pegmatite sample from zone 2 shows the upper Triassic mean age (20 spots, 235 Ma, MSWD = 2.2), while zone 3 shows the upper Ordovician mean age (17 spots, 451 Ma, MSWD = 1.5). To ensure consistency, monazite dating of two samples from Yamrang Pegmatite was done. Monazite crystals were subhedral to euhedral with dark and light patches in the BSE image with an average size of around 100 μm. The zone 1 sample shows the mean age of the middle Miocene (36 spots, 12 Ma, MSWD = 2.9), and zone 5 shows the mean age of the late Miocene (23 spots, 11 Ma, MSWD = 2.6). Hence, we concluded the crystallization age of Yamrang Pegmatite and thus the Ikhabu Pegmatite Field to be of middle to late Miocene (11-13 Ma), which is consistent with the geology of the Ikhabu Pegmatite Field.

Mineralogy of Selected Samples from Hicks Dome, Hardin County, Illinois

Ann M. Hagni[1], Laurence M. Nuelle[2]
1. Ann Hagni Consulting, LLC, Rolla, MO, USA, 2. Hicks Dome, LLC, Lutherville, MD, USA

Hicks Dome is a deep fluorite-barite deposit in the Illinois-Kentucky fluorspar district in Hardin County, Illinois. This is a mineralogical study, conducted in 2010, includes X-ray diffraction (XRD), reflected light microscopy (RLM), transmitted light microscopy (TLM), scanning electron microscopy (SEM), and energy dispersive spectroscopy (EDS) of four drill core samples. These samples are composed primarily of 10-50% fluorite, 5-15% Sr-bearing barite, 10-25% fluorapatite, 0-25% dolomite, 5-20% quartz, 10-15% clays (primarily illite), 3-7% Nb-bearing brookite, and 1-4% Nb-bearing anatase. Pyrochlore is preset in amounts up to 4% pyrochlore, along with 0-1% xenotime, and 0-2% brabantite. Trace amounts of Ba-bearing celestite, calcite, pyrite, chalcopyrite, galena, and possibly britholite are also present. Some of the fluorapatite is Y-bearing and rare earth element (REE)-bearing, and some clays are Y-bearing. Large grains of fluorite are free of inclusions, some fluorite is rimmed by fluorapatite, and much of the fluorapatite is intergrown with fluorite. Fluorapatite is very fine-grained and much of it is intergrown with fluorite. Most barite is free of inclusions, except for large grains that have been partially or mostly replaced by fluorite and/or dolomite that contain chalcopyrite and pyrite inclusions. Brookite and anatase are Nb-bearing, subhedral gains. Pyrochlore occurs as very fine-grained prismatic crystals. Xenotime also is very fine-grained. When present, brabantite rims xenotime. REE are critical to energy as they are essential for permanent magnets in wind turbines and EV motors. Y is used as an additive in allows to increase the strength of Al and Mg alloys, in microwave filters for radar, and as a catalyst. In addition to having superconductor properties, Nb is used to strengthen stainless steel alloys in jet engines and rockets, oil rigs, and oil and gas pipelines. Due to its strength and light weight, Ti is used in nuclear submarines, atomic energy icebreakers, and aviation equipment.

Alteration Mineralogy of the Mapula Porphyry Copper Prospect in the Masara Gold District, Eastern Mindanao, Philippines

Juan Alex Vianne D. Amoroso[1], Jillian Aira S. Gabo-Ratio[1], Ma. Yna Rose D. Garcia[1], Marco Alfredo Barrientos[1], Darwin Edmund L. Riguer[2], Ruel C. Mendaje[2]
1. National Institute of Geological Sciences, University of the Philippines, Diliman, Quezon City, Quezon City, Philippines, 2. Apex Mining Company Incorporated, Davao de Oro, Philippines

Southeastern Mindanao in the Philippines hosts several porphyry deposits, such as the Boyongan-Bayugo, Pantukan and Tampakan. Within this region is the Masara Gold District which has been a major source of gold with decades of exploration activities and numerous academic works. However, porphyry copper prospects were also identified in the area, but with very little detailed information available. The prospects are part of an emerging belt of intrusion-centered hydrothermal deposits in Southeastern Mindanao. This study aims to characterize the mineralization and alteration of the Mapula porphyry copper prospect which is the largest porphyry copper prospect in the Masara Gold District, which formed in association with the Masara Intrusive Complex. Diorite and diorite porphyry host the Mapula porphyry copper prospect.

At the Mapula creek, surface manifestations of the Mapula porphyry copper prospect include oxidized cross-cutting A, B, and D veins with supergene minerals malachite, azurite, and chalcocite. Based on the reconnaissance mapping and core logging of two available drill holes, five alteration types were classified: (1) propylitic alteration typified by epidote-chlorite-calcite which replaces mafic minerals. It is weakly developed distal from the deposit center; (2) argillic alteration consisting of kaolinite and illite serving as pervasive surface cover and present in the first 100m of the drillholes; (3) chlorite-sericite alteration exhibited by chlorite-pyrite with dense stockwork B veins. Bornite and chalcopyrite occur as inclusions within the porphyry veins and as disseminations in the diorite and diorite porphyry; (4) sericitic alteration typified by sericite-pyrite with dense sub-parallel B veins. Sulfide centerline of B veins consists of chalcopyrite-bornite-pyrite; and (5) potassic alteration in the diorite porphyry typified by magnetite-secondary biotite-K-feldspar with minor occurrences of M-veins. Associated ore minerals are bornite-chalcopyrite-pyrite.

With the promising alteration footprints of the Mapula porphyry copper prospect, further drilling activities would be able to delineate the porphyry copper ore body.

Skarn Mineralization in the Masara Gold District, Eastern Mindanao, Philippines

Ma Yna Rose Garcia[1], Jillian Aira Gabo-Ratio[1], Juan Alex Vianne Amoroso[1], Marco Alfredo Barrientos[1], Darwin Riguer[2], Ruel Jr. Mendaje[2]
1. University of the Philippines Diliman, Quezon City, Philippines, 2. Apex Mining Company Incorporated, Pasig City, Philippines

The Masara Gold District is located in Eastern Mindanao, Philippines. Three mineralization styles are identified in the district: the major epithermal gold-silver mineralization currently being exploited by Apex Mining Co. Inc. (AMCI), the porphyry copper-gold mineralization prospects and reported minor skarn mineralization. However, no detailed studies have been conducted on the skarn deposit and its mineralization style and characteristics, as well as the specific skarn type. This study aims to characterize and provide information on the sulfide mineralogy and paragenesis of the skarn mineralization in the Masara Gold District from detailed investigation of drill cores and underground exposures.

Initial findings of the study revealed that the calcic skarn deposit in the Masara Gold District is hosted in the Eocene Masara Formation volcaniclastic wall rock. Various overprinting of skarn zones is observed between the identified garnet-pyroxene skarn, epidote skarn and magnetite skarn. The massive garnet-pyroxene skarn is typified by dark red garnet and light green pyroxene minerals. The magnetite zone is dominantly composed of magnetite occurring as veins and massive disseminations which cut the earlier-formed garnet and pyroxene minerals. Sulfides present in this zone are pyrite and chalcopyrite, with minor sphalerite, galena, ilmenite and hematite. The epidote skarn consists mainly of retrograde yellow-green epidote overprinting the garnet-pyroxene and magnetite skarn zones. White marble is also observed containing fractures filled with dark gray Fe-Mn oxide coatings representing distal alteration of fluid escape structures. Thin garnet veins from the massive skarn are also filling the fractures. The skarn zones are being cut by the Middle Miocene Alipao Andesite.

Based from the mineralogy and dominant sulfides present, the skarn deposit in the Masara Gold District is a calcic Fe-Cu skarn.

Exploration of Buried Li Brine Resources in Andean Salars, an Environmental-friendly Option for the Long-term Sustainable Production of Lithium Brine Resources

Héctor S. Sanchez Rioja[1], Michel Lopez[2]
1. Universidad Católica de Salta, Salta, Argentina, 2. Geosciences Montpellier, University of Montpellier, France, Montpellier, France

The extraordinary global demand of Lithium has led many exploration mining companies to set their eyes on the salt lake brine Andean Plateau of Argentina, Chile and Bolivia. All these countries conform the famous Lithium triangle, that holds more than half of the total Lithium global resources. Exploration in the last decade in the Andean salars focused mainly on shallow lithium brine aquifers, which were associated with low-exploration expenditures. However, shallow brine aquifers are frequently surrounded by alluvial sediments that contain fresh and brackish water and conform breakable systems where the lithium brine body is usually in contact with fresh and brackish aquifers. The lack of effective control over the exploitation of shallow aquifers could lead to over-pumping with a risk of dilution of the lithium reservoir, provoking the affectation of the freshwater resources and sensitive ecosystems located on the margin of the salars.

Evidence from exploration works in Andean salars show deep Li brine reservoirs that include the deeper zones of the alluvial fans saturated with brine. In these areas, the mixing and freshwater zones, characteristic of shallow domains, are absent. Thus, deep reservoirs are more favorable for long-term exploitation, since the problems linked to the saline interface in the margin of the salar and its sensitive ecosystems are avoided.

A Machine Learning-based Framework for Prospectivity Mapping of Critical Minerals

Ehsan Farahbakhsh, Ralph D. Muller
EarthByte Group, School of Geosciences, University of Sydney, NSW 2006, Australia, Sydney, NSW, Australia

There has been a significant decline in the discovery of critical mineral deposits over the last two decades, and exploration geologists have begun to apply new technologies to increase the success rate of exploration projects. One of these technologies is artificial intelligence, particularly machine learning, which has been successfully applied in different fields to address a variety of problems. A range of methods has been developed for mineral prospectivity mapping. With the increase in computer power, the application of machine learning for providing accurate prospectivity maps has gained traction for exploration geologists to take full advantage of numerous datasets. This study proposes a machine learning-based framework for generating prospectivity maps of critical minerals. This framework benefits from different machine learning methods for various purposes, including an improved generative adversarial network to overcome the class imbalance problem of the training dataset, an autoencoder neural network to reduce the dimension of feature space, and the combination of a positive and unlabelled learning method and random forest as the main classifier for predicting probability values at target points. We evaluated the efficiency of our proposed framework by creating a prospectivity map of mafic-ultramafic intrusion-hosted nickel mineralization in the Gawler craton, South Australia. Various exploration datasets were used as input features, including geological, geochemical, and geophysical data. We used known mineral occurrences as positive samples and created different sets of random samples throughout the study area as unlabelled samples. Based on our results and different evaluation metrics, the model's performance is stable, and its accuracy is higher than the map generated by a conventional approach using a standard random forest classifier. Our prospectivity map shows a strong spatial correlation between high probability values and known mineral occurrences and predicts a number of potential greenfield regions with as yet undiscovered deposits.

Alteration-mineralization and Whole Rock Geochemistry at the Deep Mill Level Zone (DMLZ) Ertsberg District, Papua

Utreck F. Rumbiak
Padjadjaran University, Bandung, Jawa barat, Indonesia

The Ertsberg-Grasberg district hosts the world's largest Cu-Au (-Mo) skarn system. The mineralization is related to the Pliocene Ertsberg- and Grasberg intrusive complexes, which intruded the Jurassic-Cretaceous Kembelangan Group siliciclastics and Cenozoic New Guinea Limestone Group.

The Ertsberg granitoids have medium SiO2 and high Al2O3 contents, and consist mainly of (quartz-) monzonite and minor (monzo)diorite. Least-altered samples have upper crust-normalized enrichments in certain large ion lithophile elements (LILEs, e.g., U and K) and depletions in all high field strength elements (HFSEs, e.g., Ti, Y, and Yb). The samples are fresh to moderately-altered, featured by garnet-diopside skarn and/or porphyry-style (potassic, propylitic, phyllic) alterations and anhydrite replacement. Elemental mass balance calculation indicates that the porphyry-style alteration zones are featured by feldspar destruction and the accompanying alkali depletions, although the loss of K– Rb (in potassic and phyllic zones) and Ca-Ba-Sr (in propylitic zone) are balanced by their respective capturing in secondary K-feldspar-mica and actinolite-epidote-anhydrite. Decarbonation of the dolomitic wallrocks likely released Mg-Ca-Sr, which facilitated the subsequent forsterite-diopside exoskarn and anhydrite alteration. The enrichments of various ore-related target/pathfinder elements in these altered samples are consistent with their mineralized nature, e.g., Fe– Mn (magnetite), Cu-Fe-Bi-Se-Te (bornite), Mo-Re (molybdenite), Au-Ag-As-Bi-Te (auriferous pyrite/arsenopyrite), Zn-Fe-Mn-Sn-In (sphalerite), and Pb- Ag-Bi (galena).

Ores at Ertsberg are mainly skarn-hosted, whose formation comprises five stages: (I) prograde skarn; (II) retrograde alteration; (III) massive anhydrite replacement; (IV) late hydrothermal veining; (V) supergene alteration. Massive magnetite and main Cu– Au mineralization occurred in Stage II and III, respectively, although minor Cu mineralization persisted to Stage IV.

The Source of Mineralization in Stratabound V Rich Deposits in the Yukon Territory, Canada: Evidence from V Isotopes

Daniel D. Gregory[1], Stephan Schuth[2], Daryll B. Concepcion[1], Ashley N. Martin[3], Daixi Zhang[1]
1. University of Toronto, Toronto, ON, Canada, 2. Institute for Geology, Mineralogy, and Geophysics, Ruhr University, Bochum, Germany, 3. Institute of Mineralogy, Leibniz University, Hannover, Germany

Vanadium is becoming increasingly more important, both for its use in steel needed to build new infrastructure but also for its use in long-life batteries. As such, new sources of V need to be discovered and exploited. A potential source is highly metalliferous shales. One of the highest grade locations is in northwest Canada, where it has been found at concentrations of approximately 0.5% over tens of metres. However, it is not clear how these form so it is difficult to develop strategies for finding larger or higher grade deposits. One of the more popular mechanisms proposed is direct precipitation from seawater, however it is not clear how such high concentrations of V could be maintained for so long if reductive trapping of V from seawater were the only source. Hydrothermal fluids are another potential source but to date no known vent site has been found near these systems.

V isotopes, measured as $\delta^{51}V$, offer a chance to better elucidate the source of V as V deposited in different depositional settings and hydrothermal V all have different $\delta^{51}V$ signatures. In the modern environment $\delta^{51}V$ is -0.15 +/-0.15‰ for hydrothermal sediments, -0.9 +/-0.1‰ for oxic sediments, -0.5 +/-0.1 ‰ for anoxic sediments and -0.2 +/-0.1 ‰ for euxinic sediments. Our results range from 0.11 to -0.33 ‰ $\delta^{51}V$. While a few samples are slightly heavier than the range expected for a hydrothermal source there is no alternative probable source with a positive $\delta^{51}V$ so this is probably evidence of a minor amount of closed system behaviour. Further, while some of our results could be explained by euxinic deposition this is unlikely as there are large concentrations of barite found with the V rich horizons, something unlikely to occur in euxinic settings where all sulfate has been reduced to sulfide.

Field Observations of Magmatic Silica Caps in Over-Pressured Felsic Cupolas: Evidence for the Magmatic to Hydrothermal Transition

Douglas Kirwin, Reimar J. Seltmann
consultant, pathumtanee, Thailand

Diagnostic textures related to over-pressured volatiles and fluid saturation observed in felsic cupolas such as unidirectional solidifications textures (UST's), stockscheider, plumose feldspars and miarolitic cavities are relatively common and well-described in the literature. Field observations indicate that massive accumulations of magma-derived silica may also be present at the roof zones of cupolas forming a "silica cap". Furthermore, those silica-enriched zones (intergranular space, discrete layers, sills and dykes, massive bodies) may transition to an underlying comb quartz UST horizon reflecting magma quench processes due to undercooling and oscillatory crystallization of excess silica during the magmatic to hydrothermal transition. These do not appear to be well documented in the literature, perhaps due to lack of recognition or destruction by overprinting late-magmatic or hydrothermal events. Silica caps have been recorded in a variety of felsic cupolas, some of which are associated with intrusion-hosted mineral deposits, including porphyry copper systems. These include REE deposits at Jabal Umm Al Suqian in Saudi Arabia, the giant tungsten deposit at Panasquera in Portugal, the tin-copper porphyries in the Erzgebirge in Germany, intrusion-related gold deposits at Tsagaan Chuluut in Mongolia and Ravenswood in Australia. Examples for porphyry copper systems are Maria in Mexico, Didipio in the Philippines, Briggs in Australia and Tsagaan Survaga, Khongor and possibly Oyu Tolgoi in Mongolia, and Saginaw Hill in Arizona. In some cases, where the volatile over-pressuring continues post the deposition of the silica cap then hydrothermal breccias with barren magmatic silica clasts and matrix sulfides may form; good examples being the quartz breccias at Maria and Didipio. While most cupola-related magmatic silica caps pre-date sulfide mineralization, together with outcropping UST's zones, they are useful exploration vectors for porphyry copper and intrusion-related mineral systems. In addition, at the district scale, they provide clues as to levels of erosion within intrusion complexes.

FAMOS Insights into the Magmatic Plumbing Systems that Control the Genesis of Porphyry Copper Deposits

Jamie J. Wilkinson[1], Chetan Nathwani[1, 2], Emily Brugge[1, 2], Tom Matthews[1, 2], Simon Large[1], Matthew Loader[1], Chiara Petrone[1], David A. Holwell[3], Dan Smith[3], Jon Blundy[4], Elena Melekhova[4], Frances Jenner[5], Barbara Kunz[5], Matt Jackson[2], Haiyang Hu[2], Iain McDonald[6], Johan Lissenberg[6], Damaris Butters[7], Simon R. Tapster[8], Ben J. Williamson[9], Lawrence C. Carter[9]
1. Natural History Museum, London, United Kingdom, 2. Imperial College London, London, United Kingdom, 3. University of Leicester, Leicester, United Kingdom, 4. University of Oxford, Oxford, United Kingdom, 5. Open University, Milton Keynes, United Kingdom, 6. University of Cardiff, Cardiff, United Kingdom, 7. University of Bristol, Bristol, United Kingdom, 8. British Geological Survey, Nottingham, United Kingdom, 9. Camborne School of Mines, Corwall, United Kingdom

The FAMOS research project is a multi-million pound collaboration funded by the Natural Environment Research Council in the UK. Using a Mineral Systems approach, the project focuses on the operation of the magmatic plumbing system that permits the divergence of igneous evolution away from 'normal' arc magmatism and into a 'fertile' state from which porphyry deposits may form. Five key process 'gates' are interrogated to establish their importance in enabling or impeding the development of porphyry fertility: (i) mantle source constraints; (ii) the lower crustal 'sulfide trap'; (iii) the importance of deep-to-mid-crustal volatile saturation; (iv) the mafic trigger; and (v) extraction and focussing of fluids. Key outcomes from these studies are presented and the proxies – in the form of rock and mineral composition records – that record essential steps and which can form the basis of new exploration tools are highlighted.

Mineral Zoning in Greisen Systems: Temporal and Spatial Evolution of the Sadisdorf Greisen and Vein-style Sn-W-Li-Cu Prospect, Germany

Dino Leopardi[1, 2], Jens Gutzmer[1, 2], Bernd Lehmann[3], Mathias Burisch[1, 2]
1. Mineral Systems Analysis Group, Institut für Mineralogie, Technische Universität Bergakademie Freiberg, Freiberg, Germany, 2. Helmholtz-Zentrum Dresden-Rossendorf, Helmholtz Institute Freiberg for Resource Technology, Freiberg, Germany, 3. Mineral Resources, Technical University of Clausthal, Clausthal, Germany

The Sadisdorf Sn-W-Li-Cu prospect, Germany, hosts magmatic-hydrothermal greisen and vein-style mineralization associated to a composite granite porphyry. Here, we present petrographic and fluid inclusion data from 14 drill core intercepts covering the proximal and distal portions of the deposit. Based on this data, we aim to understand the temporal and spatial evolution of the system in order to constrain critical factors that control ore formation.

Four main mineral associations have been identified from proximal to distal with respect to the intrusion: 1) oxide, 2) oxide-sulfide, 3) sulfide and 4) fluorite. The oxide association is characterized by wolframite-cassiterite-quartz-topaz-Li-mica veins and greisen. Quartz-mica veins hosting cassiterite and/or polymetallic sulfides constitute the oxide-sulfide and sulfide associations, respectively. The fluorite assemblage consists of late fluorite veinlets and cavity-infill post-dating all previous mineral associations.

Primary fluid inclusions in quartz, topaz, cassiterite and sphalerite indicate the presence of two fluid types: a) high salinity (15-27 wt% NaCl eq.) and high homogenization temperature (Th; 270-300°C) fluid restricted to the oxide mineral association, and b) low salinity (0-13 wt% NaCl eq.) aqueous to aquo-carbonic fluid with Th ranging between 202 and 420°C in the oxide, oxide-sulfide and sulfide association, whereas maximum Th is invariably lower for the fluorite assemblage (<269°C).

Despite the overall temporal cooling trend of the system, indicated by lower Th in late fluorite, the rather constant Th of proximal and distal equivalents of oxide and sulfide assemblages does not support significant cooling along the fluid flow path during the early stage of the system along the lateral profile studied. This observation implies that the lateral mineral zoning is not a direct consequence of cooling. Instead, compositional changes (e.g., pH or fO_2) along the fluid flow path, related to e.g., water-rock interaction and/or degassing, seem to be the underlying processes that cause the observed mineral zoning.

Potentially the Largest Unexploited 'Invisible Gold' Reserve Hosted in the Detrital Pyrites of Historical Witwatersrand Tailings Dumps

Steve J. Chingwaru
Stellenbosch University, Stellenbosch, South Africa

The Witwatersrand Basin is the world's largest gold province hosting over 53 000 tons of native Au predominantly in quartz pebble conglomerates. The Witwatersrand gold tailings dumps are subjected to secondary mining operations which involve traditional extraction techniques (further comminution and cyanide leaching) that lead to 30-50 % recovery of gold missed by the historical beneficiation. This leaves a projected 1325-1855 tons of refractory gold together with an estimated 30 million tons of sulphide material which reports to the discard stream after secondary mining. The mineralogical deportment of this remaining refractory gold is not well constrained. The study is a dedicated gold deportment and ore characterization of 47 kg composite Witwatersrand tailings material from the Klerksdorp and Carletonville Goldfields. The tailings material is subjected to a modified preconcentration to define density fractions. The analysis involves optical microscopy, X-ray powder diffraction, quantitative evaluation of materials by scanning electron microscope, fire assay, aqua regia digestion, inductively coupled plasma mass spectrometry and laser ablation inductively coupled plasma mass spectrometry. Results from the analyses indicate that about 80,16 % (2,42 g/ton) of gold is distributed in the heavy mineral concentrates which are predominantly made up of silicates (44,15 %), sulphides (35,12 %), and oxides (20,12 %). Detailed in situ analyses suggest the gold in the concentrate fraction is 'invisible-' or 'solid-solution' gold hosted predominantly in pyrite and arsenian pyrites, Au grades range from 0,11- 48,69 g/ton. Given that the Witwatersrand Basin is historically a native gold deposit, identification of invisible gold in detrital pyrites (stable under Archean surface conditions) represents a potentially under-exploited resource which explains the refractory nature of the 50-70 % Au remaining after secondary mining. The implications of these findings for the economics, tailings remaining, and for the genesis of the Witwatersrand goldfields will be discussed.

The Russia–Ukraine Conflict; Understanding Short- and Long-term Impacts on Metal and Mineral Supply Chains and Prices

Simon M. Jowitt, Brian A. McNulty
University of Nevada Las Vegas, Las Vegas, NV, USA

The Russian invasion of Ukraine has already impacted global supply chains and prices of a variety of commodities, including precious and base metals and critical minerals. These impacts are likely to be long-term given that the sanctions and other actions affecting the Russian economy and the rebuilding of a post-conflict Ukraine will both persist beyond resolution of the conflict. Early impacts include the recent spikes in gold, nickel, copper, and platinum group element prices even without Russian exports of some of these commodities being sanctioned at the time of writing. This reflects the importance of both Russia and Ukraine to the global minerals industry, with Russian being a key producer of aluminum, diamonds, antimony, cobalt, gallium, germanium, magnesium, platinum group elements, silicon, vanadium, gold, rare earth elements, tellurium, titanium, tungsten, nickel, selenium, graphite, silver, copper and more. Combine this with the fact that Ukraine produces significant quantities of rutile, titanium, graphite, iron, manganese, kaolin, ilmenite, and ammonia for fertilizer and that the disruption of supply chains related to this conflict extends significantly beyond Europe means that the impact of this event is and will continue to be global. Direct and indirect short-term effects of this conflict are likely to be supply chain-related, with price spikes reflecting shortages of commodities or impeded supply chains. Longer-term effects will be dependent on the longevity of the conflict as well as the length of sanctions and the subsequent influence on the supply of minerals and metals considered key for the transition to low- and zero-CO_2 energy generation, storage and transport. Restrictions in the supply of these elements could significantly slow climate change mitigation, potentially with global consequences. This presentation will outline the effect of this conflict to date as well as scenarios for the longer-term impacts on mineral and metal supply chains.

SEG 2022 Conference: Minerals For Our Future

Global Potash Supply Chain Impacts: Geopolitical Events and Climate Change

Mark D. Cocker
USGS, Tucson, AZ, USA

During the past 20 years, the global requirement for potash as a major component of fertilizer has increased due to a growing and wealthier world population. Potash production has generally met or exceeded global demand with much of the potash exported from Canada (30%), Belarus (17%) and Russia (20%) to Brazil, India and China. The global potash supply chain is susceptible to physical disruption because potash is a perishable bulk commodity transported via specialized railcars and ships. Canadian potash is shipped 2,000 km from the Elk Point Basin to Vancouver, B.C., Thunder Bay, Ont., or Portland, OR. Belarussian potash is shipped 1,000 km via Ventspils, Latvia or Klaipeda, Lithuania on the Baltic Sea or 2,000 km to Nikolayev, Ukraine on the Black Sea, and Russian potash is shipped 2,800 to 3,000 km either via St. Petersburg or Nikolayev or 8,000 km to Vostochny on the Pacific Ocean. In 2021, EU and US sanctions on Belarussian potash caused an increase in prices from $250/t into the $600/t range. In 2022, global market potash prices have climbed to over $800/t as result of sanctions on Belarus and Russia, war-related closure of rail lines through Ukraine, and destruction of Black Sea ports effectively removing 30 to 40% of potash from global supply.

Climate change related events such as droughts across northern Europe, Russia, and Brazil and flooding in China and Canada may have more subtle but profound impacts on the potash supply chain. These events influence potash production and transportation, planting schedules, and harvests in many countries contributing to increased grain prices (up 36 to 82 percent in 2022) and restricted food supplies. A 2018 drought in Germany halted production and 7 potash train derailments in Canada during the past 7 years have disrupted rail service and impacted potash exports.

Hidden Byproduct Critical Element Supply Within Existing Mining Value Chains: Assessing the Tellurium Potential of the Australian Base and Precious Metals Sector

Brian A. McNulty, Simon M. Jowitt
University of Nevada Las Vegas, Las Vegas, NV, USA

The transition to low CO_2 energy generation, storage and transport will require a range of metals that are generally considered critical and are produced as byproducts of other metals, making increased production problematic. Here we demonstrate the hidden supply of critical elements that are necessary for the energy transition within existing mining supply chains using byproduct proxies and materials flows. The proxy-based tellurium supply potential of 52 active mines in Australia were assessed during this study. Although the abundance of tellurium and other critical elements in mineral deposits is a function of geology, the techniques used to extract the main metals (i.e., Au, Cu etc.) also influence the recoverability of byproducts metals. Tellurium materials flows were evaluated using supply potential proxies integrated with mineral processing and metal extraction methods for each mine. This analysis indicates that 23 precious metal mines have a combined Te-proxy inventory of ~31,830 t or 309 t/yr Te production potential in gold cyanide tank residuals. A further 27 base metal mines have ~57,940 t or ~204 t/yr Te production potential in sulfide metal concentrates produced by froth flotation. Australia produced an unknown amount of the ~490 t of worldwide tellurium production in 2020 with 90% of this global production from copper byproducts. This means that the Australian gold mining sector alone has the potential to produce ~63% of current annual tellurium production worldwide. However, tellurium from the Australian gold sector remains unevaluated, meaning that this potential Te supply will report to metallurgical and mine waste. The addition of 300 t/yr Te production is worth ~$21 mUSD or ~4.5 GW of power generation capacity from CdTe photovoltaic cells. Realization of this hidden tellurium supply, and other critical elements, requires a holistic approach to byproduct production and the removal of silos between resource definition, mining, smelting and refining.

A Mineralogical and Geochemical Study of Energy Critical Metals in the Carbonate-Hosted Zn-Pb Mine Tailings, Ireland

Lingli Zhou[1, 2, 3], Murray W. Hitzman[1, 3], Paul Guyett[4], Maurice Brodbeck[1, 4], John Güven[1, 3]
1. Irish Centre for Research in Applied Geosciences (iCRAG), Dublin, Ireland, 2. Geological Survey Ireland, Dublin, Ireland, 3. School of Earth Sciences, University College Dublin, Dublin, Ireland, 4. Trinity College Dublin, Dublin, Ireland

Large volumes of mining tailings have been generated as a consequence of extracting metals and minerals. Repurposing mine tailings thus becomes an essential part of sustainable mining initiatives. This work investigates the mineralogy and geochemistry of mine tailings from the historic Tynagh carbonate-hosted Zn-Pb-Ag-Cu deposit in Ireland, with a focus on understanding the deportment of energy critical metals such as Ge, Ga and In for recovery purpose. Bulk geochemical data of tailing samples have revealed areas in the tailing ponds enriched in Zn and Pb. *In-situ* SEM-EDS analysis indicates the tails contain a variety of major Zn-containing mineral phases, including a biogenic Zn mineral (>90 wt.% Zn), Zn silicate (i.e. hemimorphite), Zn sulfide (i.e. sphalerite), Zn (hydro)carbonate (i.e. hydrozincite and smithsonite), and Zn (hydro)oxide. *In-situ* LA-ICPMS analysis suggests that there is a selective mineralogical deportment of trace metals, including Ge, Ga and In. The primary sphalerite in the Tynagh deposit is relatively barren in Ge and Ga and extremely depleted in In. Germanium resides almost equally in the Zn- silicate and sulfide mineral phases which have low Fe contents (<1 wt.%). This may suggest a comparable chalcophile and lithophile affinities of Ge when Fe is not a major constituent of host mineral. In the secondary mineral phases containing higher amounts of Fe, preferred uptake of Ge is observed to be associated with enrichment of As. Gallium shows occasionally high content in couple with Cu and Ag enrichments in the Zn silicate relative to other mineral phases, reflecting a favored incorporation of Ga^{3+} through coupled substitution with monovalent cations (e.g. Cu^{+}, Ag^{+}) for Zn^{2+} in the hemimorphite in the tailings. Indium remains depleted in almost all the secondary mineral phases, which may be due to a primary depleted source and/or an oxidic and low-temperature environment unfavorable for the precipitation of In.

Bauxite Red Mud Residue as a Mineral Resource

Floyd Gray[1], Mark D. Cocker[1], Graham W. Lederer[2]
1. U.S. Geological Survey, Tucson, AZ, USA, 2. U.S. Geological Survey, Reston, VA, USA

Refining bauxite to produce alumina for aluminum smelting feedstock yields a caustic red mud residue. Globally, over 100 bauxite refineries produce about 117 Mt to 187 Mt of red mud annually. Through 2011, about 2.5 Bt of red mud has accumulated in various storage sites. Red mud residue from refineries in 25 countries has a mean composition of about 19%Al_2O_3 (), 38% Fe_2O_3 (, and 8.4%TiO_2 all of which are recovered in the aluminum smelting process. Estimated global tonnages of these oxides are approximately 570, 1,140, and 240 Mt respectively. Critical minerals such as Ga, Sc and REE are concentrated in red mud residue. Globally, mean concentrations are 107 ppm Ga,131 ppm Sc, and 1,805 ppm TREE. Estimated tonnages are on the order of 3.2 Mt Ga, 3.9 Mt Sc, and 54 Mt REE.

Historically, Greater Antilles bauxite was processed at Grammercy, LA, Corpus Christie, TX, and at five plants in Jamaica. Recently, Greater Antilles bauxite was being shipped to refineries in Canada, China, Ukraine and Russia. Red mud residue from two refineries in Jamaica contained 23–27 % Al_2O_3, 9–12 % Fe_2O_3, and 1.4–1.7 % TiO_2 as well as 1,900 to 2,600 ppm TREE and 130 to 170 ppm Sc.

While the major oxide content of red mud is relatively uniform, the critical metal content may be more variable and depend on the bauxite source. Recovery of critical elements has encountered a variety of economic and technical obstacles, but recently, several companies (Rio Tinto, Noranda, UC RUSAL) have initiated recovery of Sc from the red mud residue at their refineries.

The vast bulk of red mud residue remains an environmental problem. However, an increasing number of industrial mineral applications (over 1,200 patents) ranging from building material to pollution control and ceramics may help to alleviate environmental concerns.

Transferring Knowledge from Exploration to Minerals Processing: Innovative Use of Hyperspectral Imaging and Lithogeochemistry for Predictive Geometallurgy

Rocio T. Vargas Soto, Rick Valenta, Cathy Evans
University of Queensland, Brisbane, QLD, Australia

Ore characterization in exploration and mining is increasingly benefitting from hyperspectral imaging and multi-element geochemistry, with a promising potential to inform metallurgical process prediction across the mining value chain. However, identifying the valuable information within these large data sets is challenging, requiring the identification of quantitative geological proxies of performance distributed continuously across the orebody.

In this work, two techniques traditionally used in exploration were employed in a Copper-Gold Porphyry Deposit for comminution geometallurgical modelling. Lithogeochemistry was applied to recognize and quantify the relationship between hydrothermal alteration and comminution hardness. This helped to overcome the difficulties of recognizing and quantifying the alteration minerals which arise from subjectivity in the logging process and due to difficulties in accurately recording alteration through complex hydrothermal alteration, weathering and overprinting.

In the second approach, a workflow was developed to extract physicochemical information from the raw hyperspectral images collected with an automated core scanning system, avoiding the complexity of the conventional mineralogical interpretation process. While studies have been focused on the spectral absorption features associated with physicochemical conditions of mineral formation and proximity to ore zones, in this study, those were used to understand the links with comminution response.

Predictive models were developed using the outcomes of these approaches and the results of geometallurgical tests. They delivered strong predictors for comminution test results that outperform traditional mineralogical methods, such as automated mineralogy, X-Ray diffraction, and mineralogy interpreted from hyperspectral scanning systems.

Considering the availability and wide use of spectroscopy and geochemistry from the early stages of mining projects, this study presents an opportunity to use this information from the exploration stage for delineation of potential zones for geometallurgical sampling and process prediction.

Quantitative Assessment of Future Lithium Supply: Which Mining Projects and When?

Laura B. Andrade[1], Max Frenzel[1], Britta Bookhagen[2]
1. Helmoltz-Zentrum Dresden-Rossendorf, Institute Freiberg for Resource Technology, Freiberg, Germany,
2. Bundesanstalt für Geowissenschaften und Rohstoffe, Deutsche Rohstoff-Agentur, Berlin, Germany

Whether primary lithium (Li) sources will be able to supply the rapidly growing needs of the electric mobility transition has recently caused considerable controversy. Existing assessments are hampered by a lack of consideration for the decision-making processes occurring at the level of individual mining projects. In the present contribution, we demonstrate how these processes, as well as associated uncertainties, can be incorporated into an assessment of the likely future evolution of the global Li market.

Our method uses Monte-Carlo simulations to achieve this goal. A global database of existing Li mining projects (all development stages, including case histories) is used to build models to estimate the likelihood of each project proceeding to the next development stage in any given year, depending on specific project characteristics such as location, deposit type and ore grade, as well as market conditions, i.e., Li price and demand. Iterative stochastic simulations are then run, in which projects are moved through the development pipeline according to these estimated likelihoods, year-by-year, up to 2050. New discoveries are also included in the model, to achieve realistic results over the relatively long timespan covered by the model. Simple functional models are used to estimate future Li demand, including uncertainty. The simulations generate a large set of equally probable scenarios (1,000 or more) of which projects enter production when, whether primary supply can meet demand in any given year, and what the likely mean Li price is for that year. Summarising the data from all scenarios provides an impression of the likely evolution of the Li market up to 2050. Besides this likely market evolution, the simulation outputs can be used further to assess the probable environmental impacts of future primary Li supply.

Machine Learning – the Magic Wand for the Mining Industry?

Annelie Lundström, Mikael Arthursson
Minalyze AB, Gothenburg, Sweden

Machine learning, Artificial Intelligence and Deep Learning are methods that are influencing and changing every industry, including the mining industry. Mining companies that are not following this train now will be left behind? And those jumping on the movement will be successful? There is a risk that the answer to the questions is no, but if exploited smart it is a clear yes.

The pre-requisites for success are 1, well defined issue to handle and 2, access to high quality labelled large dataset.

The issue that this article is focused on solving through Artificial Intelligence (AI) is a well-known issue related to classification of rock-types when logging lithologies, and any geologist that have been logging rock core would be able to relate to the issue. The task of logging and classifying rock types is very time consuming, subjective, and iterative. The subjective nature of the task means that it does vary based on the person, its educational background and of course, any previous experience. A classic saying is that if four geologists look at a piece of rock, there are at least five answers to what type of rock it is. The problem could escalate in a bigger picture as there is turnover of people in the industry the new persons need to be calibrated towards the task anew which takes time and does affect the quality and time of logging.

The high-quality dataset to be used to solve the rock type classification through AI in this approach will be a combination of a wide set of data that cover both visual and compositional aspects of the rock, acquired at high-resolution, standardized to a consistent quality and correctly depth related. The machine that provides this type of combined data is the Minalyzer CS.

Automated Drillhole Target Generation for In-mine Grade Control and Out-of-Sample Resource Definition Using Site Models and Economic Constraints

Farzi Yusufali, Ilia Sucholutsky
Stratum, Toronto, ON, Canada

Producing optimally-placed drilling targets for in-mine and out-of sample exploration automatically can be valuable for reserves expansion and testing potential areas of mineralization quickly, especially when mine topography changes daily. In this paper, we propose a protocol that accepts a block model, topography map, depletion model, infrastructure files, and constraints to produce a predetermined number of drill targets that are compatible with the existing infrastructure (including underground mines) and optimized for Measured confidence with statistical safeguards included. We demonstrate this on an orogenic lode gold mine where the nugget effect is extremely high; the protocol applied to this mine resulted in over 40,000 oz of gold verified across five clusters with Measured confidence with a 2500m drilling program. Roughly 80% of drillholes hit economic mineralization and extended over 2 blocks with an average grade of almost 4g/T. Given the 60% hitrate baseline for only in-mine exploration performed by the site, this method demonstrated resilience against high grade variance in adjacent blocks and applicability to in-mine grade control and resource definition in Inferred regions of the asset.
The constraints accepted were tuned to the high grade variability within adjacent blocks, limited number of short-range drillholes, existing underground infrastructure, and physical constraints of the drill itself. Cut-off grade, maximum drillhole length, maximum angle for drill deviation, minimum spacing between drillholes, and minimum angle between the strike and drillhole, and stopes locations are also accepted as inputs. Alongside the drillhole pad location and orientation outputs, a rank is also assigned to each drillhole that quantifies the amount of verified oz per meter (in total and Measured confidence) using the reference resource model fed into the protocol.

Sequential Planning of Downhole Geophysical Surveys in Mineral Explorations by Optimizing Efficacy of Information

David Zhen Yin[1], Celine Scheidt[2], Tyler Hall[1], Lijing Wang[1], Jef Caers[1]
1. Department of Geological Sciences, Stanford University, Stanford, CA, USA, 2. Department of Energy Resources Engineering, Stanford University, Stanford, CA, USA

Downhole geophysical survey plays an important role in mineral explorations. Optimal planning of future downhole geophysical surveys is challenging. This is because the future downhole geophysical data are highly uncertain. There are no current available approaches to quantitatively assess such uncertain future geophysical data from planned boreholes. More importantly, the sequential nature of downhole geophysical acquisitions should be addressed. In this work, we introduce a new measure, named efficacy of information (EOI), to steer the sequential planning of downhole geophysical surveys. EOI quantifies how much the future actions of downhole geophysical data acquisition will reduce uncertainty in key economic parameters, such as orebody volume and depth. Maximizing EOI sequentially will lead to obtain the most informative geophysical signals to reduce the deposit exploration uncertainty. Using EOI, we will show the optimal borehole for geophysical survey changes when planning to acquire at one borehole at a time compared to acquisition at two boreholes sequentially. To emulate all the potential future geophysical data for sequential EOI calculation, we train an artificial neural network (ANN). The trained ANN can fast generate the potential borehole geophysical data once given the borehole locations, thus enhancing the computational speed. We will demonstrate this approach on a real metal deposit derived from the Mid-Continent Rift system.

Geology from Geochemistry, an Emerging Opportunity

Ross Large
University of Tasmania, Hobart, TAS, Australia

The recent common use of 4-acid ICP-MS multi-element geochemistry for soil samples, stream sediments and drill chips has been a significant breakthrough for mineral exploration. This approach now enables the geoscientist to interpret geology from geochemistry. By the use of selected trace elements and in particular trace element ratios such as Ti/Zr, Hf/Sc, V/Sc, La/Y, Sr/Y etc, combined with major element correlations, it is possible to distinguish a range of rock types from drill chips, fine drill cuttings and soils; a technique that has only recently become available. For example felsic volcanics, mafic volcanics, clastic sedimentary rocks, limestones, dolomites and black shales can be confidently recognised by the use of multi-element geochemistry, which is invaluable in cases where geological maps or interpretation from chips and fines are not readily available. Even fractionation series and hydrothermal alteration facies in a sequence of volcanic or plutonic rocks, can be recognized and used to focus exploration targeting. The lack of geological information presented by air core, reverse circulation and percussion drillng is now partially offset by using multi-element geochemistry on chips and/or fines to give a geological picture approaching that available from diamond drilling.

In this presentation I will show some case studies in environments that vary from the deserts of Western Australia to the semi-arctic conditions of Alaska and the Yukon. When this approach is combined with a second recent innovation of using LA-ICPMS pyrite chemistry to vector to ore, the knew knowledge available and its application to mineral exploration is very impressive. This approach is particularly relevant to exploring under cover where bedrock geological maps are not available.

Want to Make an Impact? Become a Production Geologist

Peter Taylor
Hecla Greens Creek Mining Company, Juneau, AK, USA

Recruiting employees with fresh skill sets and diverse perspectives is one of the greatest emerging opportunities to improve the geosciences in the mining industry. Unfortunately, many American geologists are unaware that a career in mining is even an option in the USA. The subset of young geologists that are interested in the mining industry are more likely to be exposed to exploration than mining through their academic mentors and student research projects. Among both seasoned geoscience professionals and fresh graduates, there is a lack of awareness of how important and rewarding careers in production geology can be. This presentation offers an inside look at the day-to-day life of a production geologist at Hecla Greens Creek's world class underground Ag-Au-Zn-Pb mine. Real photos, scenarios, analytical data, models, and maps are discussed to show how geologists at mines use their passion for science to make immediate positive impacts on their companies and communities. Production geologists do ore control: the highest impact step in the downstream processing of ore and environmental remediation of waste. The mapping, sampling, and insights of production geologists continuously improve a mine's geologic model. These geologic models drive the cost optimization of mining and frame the hypotheses for continued resource exploration. Additionally, as crucial liaisons between miners and managers, production geologists make an outsized business impact as safety advocates and project coordinators. Mining's greatest impacts on society are at the point of extraction. The challenges of being a mine geologist at Greens Creek illustrate why members of SEG should promote and discuss careers in production geology on an equal footing as those in academia or exploration. As an industry and a science, we will better capitalize on emerging opportunities if we attract smart people to the roles where they can make the greatest impact.

Trace Elements and U-Pb Dating of Zircon from the Humpa Leu East Porphyry Cu-Au Prospect, Sumbawa Island, Indonesia: Implications for Petrogenesis and Magma Fertility

Fadlin Fadlin[1, 2], Ryohei Takahashi[1], Andrea Agangi[1], Marlina Elburg[3], Arifudin Idrus[4], Bronto Sutopo[5], Rachmat Pratiwinda[6]
1. Akita University, Japan, Japan, 2. Jenderal Soedirman University, Purwokerto, Indonesia, 3. University of Johannesburg, Johannesburg, South Africa, 4. Universitas Gadjah Mada, Yogyakarta, Indonesia, 5. PT. Antam, Tbk, Jakarta, Indonesia, 6. PT. Sumbawa Timur Mining, Jakarta, Indonesia

The Humpa Leu East (HLE) porphyry Cu-Au prospect is located in the eastern part of the Sunda metallogenic belt, Sumbawa Island, Indonesia. The belt hosts many world-class porphyry Cu-Au deposits formed around 5.7-0.5 Ma, including Tujuh Bukit, Batu Hijau, Elang, and Onto-Hu'u. As the eastern Sunda arc is geologically young and still developing, it is difficult to find mineralization cropping out at the surface. Therefore, more advanced and cost-effective geochemical techniques are required to reduce costs and time in exploration. This study focused on the trace elements of zircon in porphyritic intrusive rocks to understand the petrogenesis and determine the magma fertility based on LA-ICP-MS analysis. This can be potentially applied in discovering new porphyry prospects in the eastern Sunda arc. The HLE zircon has a slightly high U/Yb > 0.1 with low Ti content < 20 ppm indicate that the intrusive rocks related to the mineralization in the HLE are a product of continental crust/active continental margin, reflecting the subduction of the Indian oceanic plate beneath the Sunda continental plate. Moderate pyramid and prismatic zircon shapes indicate that the HLE zircon is a product of calc-alkaline magmas and 650-800 °C crystallization temperature, respectively. Ti-in-zircon thermometry shows similar result ranging from 737 to 883 °C (av. 779 °C). Furthermore, U-Pb dating of zircon on the HLE yielded dates of 1.2 to 1.0 Ma. In this study, we observed that zircon in the intrusive rocks related to the mineralization show significant oxidized (Eu/Eu* > 0.4 and Ce/Ce* > 50) and high water content conditions as evidence of amphibole fractionation (Yb/Dy > 5 and Y/Ho > 30). In parallel with many other previous studies, geochemistry of zircon from HLE porphyry-related intrusions suggested oxidized and high water content magma which is applicable for magma fertility assessment, especially for the eastern Sunda metallogenic belt.

SEG 2022 Conference: Minerals For Our Future

In Situ Rb-Sr Dating of Hydrothermal Mineral Deposits by LA-ICP-MS/MS

Shiqiang Huang, Zhaoshan Chang
Colorado School of Mines, Golden, CO, USA

The ages of alteration and mineralization are important for both economic geology research and mineral deposit exploration. K-bearing minerals such as biotite, muscovite, illite, K-feldspar, alunite, and some amphiboles are common alteration minerals in many deposit types. Such minerals are currently dated using Ar-Ar or K-Ar methods. With the new generation triple quadrupole ICP-MS/MS, it is now possible to achieve online separation of ^{87}Rb and ^{87}Sr and thus allow in-situ Rb-Sr dating of these minerals with laser ablation sample introduction.

In this study, we used LA-ICP-MS/MS to determine Rb-Sr ages of micas from 7 hydrothermal deposits with known ages (306 Ma to 1.4 Ma), to test the capabilities of this technique. The separation of ^{87}Sr from ^{87}Rb was achieved using N_2O cell gas. The isotopic ratios were calibrated with Mica-Mg-NP and NIST 610. Uncertainties in all steps were incorporated into the final uncertainty. The results showed that the Rb-Sr dating of micas can achieve accuracy of <1% deviation from the known ages. The isochron age precision is typically ~1 to 1.5% (at two sigma level) for minerals with ages older than 40 Ma, and ~15% or worse for minerals younger than 20 Ma, which is likely because of (1) the low radiogenic Sr due to the short accumulation time, and (2) the narrow range of $^{87}Rb/^{86}Sr$ ratios in the samples. As compared to the Ar-Ar dating, this technique has shorter turnaround time and larger throughput, plus higher closure temperature (>500 °C), making the Re-Sr system more likely to survive in the later thermal events than the Ar-Ar system (closure T ~350 °C). These features make the LA-ICP-MS/MS in situ Rb-Sr dating suitable for geochronologic studies of mineral deposits.

A Study of the Diatomite Deposit in Shengzhou of Eastern Zhejiang Province, China

Xu Jia, Xiaobin Wu, Baojin Zhao, Chao Zhang
Southwest Petroleum University, Chengdu, China

The Shengzhou diatomite deposit is one of the large ones in China, which occurs in the Shengzhou-Xinchang volcanic basin in Zhejiang Province. It was formed in the Neogene. We conducted scanning electron microscopy (SEM), X-ray Diffraction (XRD) and geochemical studies on diatomite samples. Through SEM, the predominant species of diatoms are m*elosira granulata*, c*yclotella shengxianensis huang,* and n*avicula while* XRD analysis showed that the mineral components of the diatomite include predominante quartz and clays with minor amounts of feldspar, carbonates and pyrite. The clay minerals are mainly composed of illite, kaolinite, and illite/smectite mixture. Geochemical results showed that the high content of SiO_2 ranges from 59% to 70%, which was mainly related to the accumulation of diatoms. Trace metals such as Mo, U, V, Co, Ni, Cu, and Zn in the diamomite are slightly higher than the abundance values of the upper crust. There are significant layers of basaltic lava and other vocanic rocks interbeded with diatomite horizons in the area so that we believe formation of the diatomite was related to the eruption of the basaltic lava series which is an indicator of exploration for the deposit in the area and similar environment.

Contribution to the Magnesite Formation in the Tshipise Magnesite Field, Limpopo Province of South Africa

Lutendo D. Mutshaine[1], Baojin Zhao[1, 2], Thibedi Ramontja[1, 2], Lindani Ncube[1], Elna Van Niekerk[1]
1. SACNASP, Pretoria, South Africa, 2. GSSA, JHB, South Africa, 3. GSSA, Akasia, South Africa

Three open pit mines were developed in the Tshipise Magnesite Field in the Limpopo Province, South Africa, including the Folovhodwe, Venda (Venmag) and Nyala Magnesite Mines, but the host rocks of the magnesite mineralisation and its formation mechanism proved to be contradictory with previous studies. This study aims to address and solve these opposing views. The methodologies, used, included field investigation, geochemical sampling, mineralogical and geochemical studies by means of microscopy and X-ray Diffraction (XRD), and geochemistry by X-ray Fluorescence (XRF). It was ultimately concluded that the primary magnesite host and magnesium source rock is ultramafic, rather than doleritic, as previously reported. The mineralisation mechanism occurred in four phases: (1) Starting with the intrusion of ultramafic rocks; (2) Being followed by the dolerite dykes; (3) Hydrothermal fluids enriched with CO_2; and (4) Precipitation of magnesite in ultramafic bedding planes and joints.

Keywords: Magnesite, Folovhodwe Mine, Venmag Mine, Nyala Mine, Limpopo Province

Paramount Ag – Au Potential under Barren Lithocaps in Mexico's Prolific Silver Belt

Jorge E. Ordonez
Cima Yubara, S.A.P.I. de C.V., Mexico City, CMX, Mexico

Mexico's Prolific Silver Belt high grade Ag-Au veins in epithermal, XVIth Century Mining Districts, amongst the worlds largest producers, Pachuca, Zacatecas, Guanajuato, Fresnillo, etc each produced +500Million to +2Billion ounces of Ag-Au. In these old mining districts, centuries of mining of outcropping Ag-Au high grade veins led to the exploration of neighboring barren lithocaps which were found to overlie deeper, blind high-grade veins. This suggests that similar barren lithocaps with no vein outcrops in the prolific Silver Belt in Central Mexico may also have close spatial relationships with significant, deep Ag-Au mineralized vein systems. There are many unexplored prominent barren lithocaps along the trends of these old districts which have no vein outcrops; exploration of these barren lithocaps with no known vein outcrops nor previous mine workings is in its preliminary stage. However, some significant discoveries have already been confirmed. Early 2000's, Peter Megaw Mag Silver drilled veins with +700 AgEq grams at +500 metres depth at Juanicipio 8 kilometres SW of Fresnillo; this has developed into a prosperous mine. Early 2010's, Michael Clarke Plata Latina drilled +7 metres with +3,500 AgEq grams at +500 metres depth at the Naranjillo Property 35 kilometres SE of Guanajuato; the NI43-101 Report describes the successful rationale for drilling deep below the barren lithocaps with no known vein outcrops nor previous mine workings. February 2021, Electrum Group CEO Thomas Kaplan informed the WSJ of "...a mine in Mexico called Celaya that has resources of at least 800 million ounces…". There is a significant number of prominent barren lithocaps with no known vein outcrops nor previous mine workings in the Prolific Silver Belt in Central Mexico which present very attractive exploration targets with potential for significant Ag-Au mineralization. Exploration of these barren lithocaps proffers superb exploration targets for the discovery of major, blind Ag-Au resources.

Zircon Oxybarometry and Hygrometry Discriminate Porphyry-Copper Ore-Forming Magmas from Ordinary, Infertile Arc Magmas: Guides to Exploration Targets

Robert R. Loucks, Marco L. Fiorentini
Centre for Exploration Targeting, University of Western Australia, Crawley, WA, Australia

The ratio (Eu/Eu*)/Yb in zircon increases as hydration state of its parent silicate melt increases (Loucks et al., 2016). The ratio $Ce/\sqrt{(U_i \times Ti)}$ in zircon increases as oxidation state (ΔFMQ) of its parental silicate melt increases (Loucks et al., 2020). U_i is the initial ppm U in zircon, corrected for post-crystallization radioactive decay. Our database of trace-element compositions and U-Pb ages of >8,000 zircons from porphyry-copper ore-forming intrusions in 40 mining districts and from many unmineralized igneous suites at convergent plate margins demonstrate that plots of zircon (Eu/Eu*)/Yb versus $Ce/\sqrt{(U_i \times Ti)}$ are remarkably effective in discriminating zircons in copper-ore-forming granitoid intrusions from zircons in copper-infertile, ordinary magmas of calc-alkalic granitoid compositions. During fluid-undersaturated and plagioclase-undersaturated fractional crystallization of arc magmas at high pressure, the residual melt's (Eu/Eu*)/Yb and $Ce/\sqrt{(U_i \times Ti)}$ rise in tandem. As hydrous granitoid magma ascends through the crust and exsolves hydrothermal fluid in response to decompression, plagioclase crystallizes and the melt's (Eu/Eu*)/Yb ratio decreases. As hydrothermal fluid exsolves, reduced members of Fe, S, C, and H redox couples are preferentially partitioned into exsolving, gravitationally segregating hydrothermal fluid, leaving an increasingly oxidized residual silicate melt. Zircon crystals growing from decompressing, dehydrating melt record the decreasing hydration state of the melt as a typically ~10-fold decrease in (Eu/Eu*)/Yb and simultaneous increase of $Ce/\sqrt{(U_i \times Ti)}$ by a factor of ~3. We introduce a compound hygrometer-oxybarometer indicator of magmatic copper fertility, $10^4 \times (Eu/Eu^*)/Yb_N + 5 \times Ce/\sqrt{(U_i \times Ti)}$ in zircon, which compresses the range of variation to factor of only about 2 within zircon grain populations in individual granitoid plutons, but typically is 2 to 20 times higher in Cu-ore-forming intrusions than in unmineralized, infertile granitoid intrusions. These zircon trace-element discriminants of Cu-fertile arc magmas are applicable to detrital zircons for identification of watersheds containing Cu-prospective igneous complexes upstream from the sediment sampling site.

3D Locally Varying Anisotropy Fields Produced Automatically From Sparse, Continuous Numeric Point Data

Alexander M. Wilson, Ali Mohammad Mehr, Samuel F. Cantor
Minerva Intelligence, Vancouver, BC, Canada

Geological phenomena often do not conveniently fit into a stationary random function model. Structural deposit influences such as folding, faulting, and non-linear fluid interactions may result in deposit complexity that is challenging to model accurately, particularly when using statistical interpolation techniques that rely on distance-organized points as input (e.g., kriging).

We present a novel machine learning method that can automatically quantify three-dimensional, locally varying anisotropy (LVA) fields directly from unprocessed (sparse) continuous numeric point data. The method utilizes an overlapping, moving window to repeatedly sample input data subsets and applies a directional continuity mapping algorithm that can determine the optimal anisotropy ellipsoid configuration (i.e., orientation, shape, and size) for each subset area. The algorithm can manage the presence of discrete, sub-parallel features of interest and adjust for missing data and data boundaries by trending towards a conservative isotropic case. The generated LVA fields may be used to i) guide structural and geometric exploratory data analysis of complex systems, and ii) inform more realistic 3D block model estimations that can dynamically adjust the sample search neighbourhoods and predictive weightings to account for the local changes in anisotropy.

Ultimately, our method represents an important step towards a reliable and accurate semi-autonomous geologic modelling system that is capable of seamlessly capturing the nuances and complexity inherent to many geologic systems.

SEG 2022 Conference: Minerals For Our Future

Whole Rock Analysis with LIBS – Paving the Way to Downhole Chemical Investigations

Fernando Fagundes Fontana[1, 2], Ben van der Hoek[1, 2], Caroline Tiddy[1, 2], Steven Tassios[3, 2], Neil Francis[4, 2], Jessica Stromberg[4, 2], Yulia Uvarova[4, 2]
1. University of South Australia, Mawson Lakes, SA, Australia, 2. Mineral Exploration Cooperative Research Centre - MinEx CRC, Mawson Lakes, SA, Australia, 3. CSIRO Mineral Resources, Clayton, VIC, Australia, 4. CSIRO Mineral Resources, Kensington, WA, Australia

Laser-Induced Breakdown Spectroscopy (LIBS) is a well-established in situ analytical technique capable of multi-element analysis with limited to no sample preparation requirement. LIBS has been applied to downhole groundwater monitoring and analysis of geological materials in harsh and remote environments such as on Mars and in the deep sea. In this study, LIBS is adapted to geochemical analysis in a drill hole environment. The development of a downhole deployable tool for chemical analysis would represent a paradigm shift for the exploration and mining industries by enabling (near) real-time geochemical assay. Here we demonstrate successful emulation of whole rock geochemical assay for major elements in discrete rock types of various textures and mineralogy (13 rock samples in total) using LIBS spot data, collected along continuous, 1mm-spaced transects over 1-metre drill core intervals (i.e. 1000 LIBS spot analyses per metre). Results of whole rock geochemistry from traditional laboratory assay and LIBS are remarkably similar. For instance, silicon differences between results do not surpass 14% for the whole sample set. Investigation of optimal LIBS sampling strategies for whole rock analysis shows that a range in the number of LIBS spot analyses is required to emulate whole rock geochemistry within 1% error and 95% confidence (e.g. microgranite: ~50 spots; dolostone: ~560 spots per metre). The accuracy of whole rock LIBS analysis and optimal number of analyses are largely dependent upon elemental range of distribution, grain size, and element deployment (host minerals behavior). Overall, the results of this study show promise in development of strategies for rapid whole rock geochemical analysis down a drill hole.

Decision Theoretic Methods for Mineral Exploration

John Mern
KoBold Metals, Berkeley, CA, USA

As the global demand for batteries increases, the need to find to find new battery metals deposits become more critical. Exploring the subsurface for new deposits requires gathering data about prospective targets through geophysical testing, geochemical surveys, and drill sampling. Finding a prospective subsurface deposit can require a large amount of well targeted testing. Testing is often conducted according to fixed test schedules that define the sequence measurements to be taken before any measurements are taken. Decision theory suggests that dynamic plans, in which the measurements taken adapt based on what is observed in previous measurements, will outperform static plans on most information-gathering tasks. This improved performance can allow decisions to be made with fewer measurements required.

We have formulated mineral exploration problems as sequntial optimization problems called partially observable Markov decision processs (POMDPs). We develop methods to solve for optimal dynamic sequntial measurement plans using POMDP search methods. Each step of the resulting dynamic plan takes as input a distribution over the possible states of the sub-surface geology. Along with the dynamic planning process, we implemented a Baysian regression method to update an initial distribution with the information gained from each measurement. When testing a prospective target, a natural question that arises is when has enough data been gathered to falsify the hypothesis that an economic deposit exists. The distribution resulting from each update provides a quantiative estimate of the deposit state that can be used to make a principled decision of when to falsify a deposit hypothesis. We have implemented these methods for a drill measurement campaign for a prospective deposit and compared the predicted results to a baseline measurement method.

Modelling Carbon Sequestration Capacity of Ultramafic Rocks for Use as a Criteria for Critical Metal Exploration in British Columbia, Canada

Dianne Mitchinson[1], Jamie Cutts[2], Dominique Fournier[3], Annika Naylor[2], Greg Dipple[2], Craig J. Hart[4], Connor Turvey[2], Mana Rahimi[4], Dejan Milidragovic[5]
1. Mineral Deposit Research Unit, University of British Columbia, Vancouver, BC, Canada, 2. University of British Columbia, Vancouver, BC, Canada, 3. Mira Geoscience Ltd., Vancouver, BC, Canada, 4. Independent, Vancouver, BC, Canada, 5. Natural Resources Canada, Vancouver, BC, Canada

Significant quantities of battery metals need to be mined to help electrify the economy and reach net-zero carbon emissions targets by 2050. Some of these critical metals, like nickel and cobalt, are hosted in ultramafic rocks that have the capacity to naturally sequester CO_2. As such, certain mine sites could in fact operate with net zero or even net negative carbon emissions. Serpentinized ultramafic rocks are highly reactive with CO_2 - magnesium in minerals like brucite and serpentine react with CO_2 to form carbonate minerals that are stable over geologic time. This process, carbon mineralization, has been found to be occurring passively at several mines around the globe. The capability of serpentinized rocks to store CO_2 is starting to be considered as an exploration criteria by mineral exploration and mining companies who hope to eliminate mine emissions.

Serpentinization of ultramafic rocks generates magnetite, causing the rocks to become magnetically susceptible. Using magnetic susceptibility as a proxy for the extent of serpentinization of ultramafic rocks, we evaluated regional magnetic data to determine the volumes of serpentinized ultramafic rocks in British Columbia (BC), Canada. We identified ninety-nine sites in BC where mapped ultramafic units correlate with positive magnetic anomalies indicative of serpentinization. The anomalies were modelled in 3D to yield volumes from which carbon mineralization capacity per site was calculated. Considering only the near-surface bodies (<1km depth), we calculated that approximately 30 Gt CO_2 could be captured by ultramafic rocks in BC, if these rocks were crushed as tailings. This is the equivalent of >400 years of BC's CO_2 emissions. Exploration companies can use the 3D models and estimated carbon mineralization capacities alongside other exploration criteria to prioritize exploration sites based on their potential to reduce CO_2 emissions, and to plan for technology and infrastructure that could support or enhance carbon capture.

Machine Learning to Boost Hyperspectral Imaging

Cecilia Contreras[2], Richard Gloaguen[1], Pedram Ghamisi[1], Behnood Rasti[1], Sandra Lorenz[1]
1. Helmholtz Institute Freiberg for Resource Technology, Freiberg, Germany, 2. TheiaX, Freiberg, Germany

The digitization and automation of the raw material sector are required to attain the targets set by the Paris Agreements and support the sustainable development goals defined by the United Nations. While many aspects of the industry will be affected, most of the technological innovations will require smart imaging sensors. In this respect, hyperspectral imaging (HSI) is becoming prevalent in the extractive industry. The ability to image and characterize key minerals at different scales is beneficial in exploration, beneficiation, metallurgy, and monitoring (post-)mining activities. HSI is usually considered as an image or in a spectroscopic workflow (target fingerprints). We have recently developed a series of machine learning tools that can boost the impact of HSI on geometallurgy and monitoring. After proper denoising and fusion, HSI data can be classified to produce maps that can be used to support decision-making. For example, we developed efficient unsupervised approaches based on deep-learning (Auto Encoder architecture) to cluster large HSI datasets and display information in augmented reality. We also demonstrated that HSI data allow the upscaling of mineralogical and geochemical information. We used an automatic high-resolution mineralogical imaging system (i.e., scanning electron microscopy-mineral liberation analysis) to generate training labels. We adopted a soft labeling strategy and then used random forest and support vector machines to map the entire drill-cores. In another example, we used resolution enhancement as the spatial sampling of hyperspectral drill core scanning sensors does not allow rendering the textural and mineral diversity required to map minerals with low abundances or fine structures such as veins and faults. We proposed an algorithm based on coupled non-negative matrix factorization, using hyperspectral and RGB images to increase the spatial resolution of HSI data and improve mineral and structure mapping.

Title: Remote Sensing of Gudjareti- Khachkovi Gold Ore Field. Adjara-Trialeti Folded Zone, Lesser Caucasus (Georgia)

Theme: New Frontiers, Innovative Technologies, and Emerging Opportunities in Economic Geology

Presentation Type: Poster

Session:

Authors: Giorgi Mindiashvili
Ivane Javakhishvili Tbilisi state university, Tbilisi, Georgia

Abstract Text:

This research consists of interpretations of more than 40 data obtained by remote sensing analyses and techniques in order to detect geological features and potential of probable Gudjareti- Khachkovi ore fields in the area located in South-East Georgia (Adjara-Trialeti folded zone, Lesser Caucasus) using Terra ASTER Multispectral satellite data.

Technical specifications for the ASTER data used are as follows:

Granule ID: AST3A1 0409160806091107270062

Processing Level: 3

Acquisition Date: 20040916

Scene ID: [171, 87, 1]

Processed Bands: "01023N3B0405060708091011121314"

Cloud Coverage: 2

The area is quite intensive in terms of vegetation, snow, glacier, and cloudiness. Therefore, the best satellite data among all ASTER images have been ordered that could represent the area the best way and which has the least snow and cloud cover. The acquisition date of the satellite data is 24.08.2012. It is noticeable that although the time of acquisition is autumn season, excess snow covers still remain on mountain tops. The ASTER image used in this study has approximately 2% cloudiness, 20% vegetation, and 35% snow cover in the whole region. In this very highly vegetated territory, there have been detected different rock types covering 10% of the study area and lithological descriptions, mineral mapping techniques, and analysis towards determination of geological features have been made.

SEG 2022 Conference: Minerals For Our Future

Vectors to Porphyry Copper Deposits Hosted in Carbonate Rocks: an Update on the Bingham to Stockton Carbonate Transect Geochemistry

Michael J. Kirschbaum[1], Zhaoshan Chang[1], Adam Pacey[3], Paul Agnew[3], Phillip Nickerson[2], Kim Schroeder[2], Kathleen Gundy[2], Andrew Sasso[2]
1. Colorado School of Mines, Golden, CO, USA, 2. Rio Tinto, Salt Lake City, UT, USA, 3. Rio Tinto, Bundoora, VIC, Australia

Copper is one of the most important metals utilized by modern civilizations and it is estimated that ~75% of the known global copper resources (~3,100 Mt) are contained within porphyry deposits. Most future porphyry discoveries will require deep drilling of systems with only minor (or completely absent) surficial expression of ore. Researchers in the last few decades have made progress on identifying the distal geochemical footprints and vectors to porphyry copper deposits but most of these studies have focused on igneous host rocks. Our research aims to expand upon this knowledge by investigating the proximal to far-field geochemical signals of porphyry copper deposits emplaced within carbonate rocks.

In this Bingham – Stockton research update we present geochemical trends in two laterally continuous carbonate units, the Jordan and Commercial limestone beds. These two carbonate units were sampled along a 17km transect between the giant Bingham and the smaller Stockton porphyry systems. Whole rock analyses reveal several notable elemental trends:

1) Be, Cu, Fe, Ga, In, Mo, Ni, Re, S, Se, Sn, V, Pd and most REE's show higher concentrations near Bingham with a sharp drop ~1km outboard

2) Ag, As, Nb, Nd, P, and Pb decreases to a distance of ~5km from Bingham to background.

3) Ba, Cd, Mn, and Zn increase to a distance of ~2km from Bingham, then decreases outboard

4) Sr increases outward to ~2km, then stabilizes.

Drilled vein samples show similar relations, although Ba, Cd, Ga, Nb, Ni, Pd, Re, REE, Se, and V do not show trends, whereas Ge shows an increasing outwards trend. Whole rock C-O isotopes show hydrothermal signatures ($\delta 18O<15‰$, $\delta 13C<-3.3‰$) within 2km of Bingham, then jump to marine carbonate signatures ($\delta 18O$ 20-29‰, $\delta 13C$ -1.8 to 3.6‰) outwards without temperature corrections.

Hyperspectral Satellite Remote Sensing for District-scale Exploration in the Coastal Cordillera of Northern Chile: Alteration Mapping as a Vector to the Mineralization

Rita Chirico[1], Anna Sorrentino[1], Francesca Corrado[1], Carsten Laukamp[2], Nicola Mondillo[1, 3]
1. University of Naples Federico II, Naples, Italy, 2. CSIRO Mineral Resources, Kensington, Perth, WA, Australia, 3. Natural History Museum, London, United Kingdom

In this study, we propose an application of satellite hyperspectral remote sensing for identifying and mapping specific minerals associated with the surface-exposed hydrothermal and supergene alteration assemblages vectoring to ore deposits in the Coastal Cordillera of northern Chile. The area is the host of several iron, copper, and minor gold, silver, and zinc ore bodies of Early Jurassic to mid-Cretaceous age. Host rocks to the precious and base metal deposits commonly exhibit hydrothermal alteration patterns, characterized by both upward and outward zonation. Supergene alteration of the ore bodies causes oxidized iron-rich blankets. To map the relative abundances and compositions of specific supergene and hydrothermal alteration minerals, which are optically active in the Visible Near (VNIR) to Short Wave Infrared (SWIR) wavelength regions, such as Fe-oxides and hydroxides (hematite-goethite), di- and tri-octahedral phyllosilicates (e.g., micas-kaolinite-chlorite), hydroxyl-bearing sulfates (e.g., alunite) and epidote, a range of band ratios in the region around 900 nm and from 2,100 nm to 2,300 nm, were applied to PRISMA satellite imagery. PRISMA, a hyperspectral spaceborne imaging spectrometer funded by the Italian Space Agency, provides images at a spectral resolution of ~13 nm in a continuum of 234 spectral bands in the VNIR and SWIR wavelength range (400-2500 nm). Since it covers the mineral-diagnostic wavelength regions at hyperspectral resolution and signal-to-noise ratio (SNR) from ≥ 200:1 (VNIR) to >100:1 (SWIR 2), it has the potential to accurately recognize and map the earth's surface materials, discriminating between different geological targets according to their diagnostic spectral response in each of the narrow bands. The preliminary results assess that PRISMA permits reliable discrimination of alteration minerals associated with hypogene and supergene alteration bands, which may provide useful vectors to prospective mineralized areas. The method can be considered a valuable strategy for cost-effective and safe mineral exploration at regional scale.

Evaluation of Normalization Methods Applied to Short-Wavelength Infrared Spectroscopy Mineral Databases from Multiple Instruments and for Vectoring Analysis Exploration

Juan C. Paredes[1], Yan C. Trigos[1], Camilo Uribe-Mogollon[2]
1. EAFIT UNIVERSITY, Medellin, Colombia, 2. SM Energy, DENVER, CO, USA

Over the last decade, significant advances have been made with the application of short-wave infrared (SWIR) spectroscopy to detect subtle variations in the geochemistry of alteration minerals that are used as vectors for exploration of hydrothermal ore deposits. In the context of porphyry deposits, the study of the spectral variations in hydrothermal minerals such as white mica, alunite, and chlorite have provided relevant vectoring information produced by changes in temperature, pH, and/or oxidation state of the fluids towards the mineralized center. Successful application of SWIR spectroscopy as a vectoring tool is based on the instrument precision. However, during data collection most exploration projects may use multiple SWIR instruments with different factory calibration settings. As a result, in cases where no inter-instrument calibration is performed, variability of the value of a spectral feature among SWIR instruments has been identified and the application of data for vector analysis is often difficult. In the present work, we evaluate four of the most common normalization methods: 1) rescaling, 2) mean normalization, 3) Z-score, and 4) kernel density estimation applied to three white mica SWIR spectroscopy datasets collected by two different field instruments (TerraSpecTM and OreXpressTM) in samples presenting phyllic alteration from the study of the Grasshopper porphyry prospect by Uribe-Mogollon and Maher (2020). The outcome of this work is the development of a methodology to compare, quantify and correct the SWIR databases biases for an accurately integration of data from multiple instruments for vectoring analysis application.

Drones for Remote and Autonomous Multi-sensor Mapping in Mineral Exploration

Sandra Lorenz, Yuleika Madriz, René Booysen, Samuel T. Thiele, Moritz Kirsch, Richard Gloaguen
Helmholtz-Zentrum Dresden-Rossendorf, Helmholtz-Institute Freiberg for Resource Technology, Freiberg, Germany

Uncrewed aerial vehicles (UAVs), also known as drones, have become an important branch of development in the field of innovative exploration technology. Short turnaround times, the highly adaptable nature of drone platforms, and the growing variety of sensors that can be deployed are driving increased interest in implementing drone-based mapping into mineral exploration workflows. Drone-based adaptation of technologies previously applied in airborne or ground-based campaigns now enables rapid mapping of geologic targets in unprecedented detail. The ability to objectively map topography and surface mineral composition using imaging sensors (including RGB, multi- and hyperspectral cameras) and subsurface physical properties using geophysical sensors such as magnetics and radiometrics has shown to add impressive value to conventional mapping workflows. Regulatory and technical barriers, however, often force a difficult trade-off between sensor payload and flight time on drone-based surveys. In an applicational field, where short turnaround times are key and targets of interest are remote and difficult to access, this often results in the use of light-weight sensors and single-sensor acquisitions.

In this contribution, we demonstrate the challenges and opportunities provided by multi-modal drone-based data in the framework of mineral exploration. In several case studies, we showcase the added value of integrating surface (spectral imaging) and subsurface (geophysical) data for better target characterization. We finally give an outlook on autonomous and multi-drone data acquisition for a targeted and more efficient characterization.

Automated Detection of Geophysical Features Characteristic of Carbonatites-Aalkaline Ring Complexes

Malcolm Aranha[1], Ushashi Chaudhuri[1], Alok Porwal[1, 2]
1. Indian Institute of Technology Bombay, MUMBAI, MH, India, 2. Centre for Exploration Targeting, University of Western Australia, Crawley, WA, Australia

Carbonatite-alkaline complexes are rare rock types that are often enriched in critical elements such as rare earth elements (REE), U, Th, Nb, and sometimes Au, Cu, PGE, and diamonds. These rocks typically occur as ring complexes or linear dykes in extensional settings such as intra-continental rifts associated with deep crustal faults and show ring-like oval-circular or linear geometries. These rocks often display well-defined anomaly patterns in geophysical datasets, which can map poorly exposed carbonatite-alkaline complexes in deeply weathered and under-cover terranes. The geophysical anomalies are generally manually interpreted and analysed using visual methods, which are time-consuming and often subjective. This study develops a machine learning-based, completely automated workflow incorporating convolutional neural networks (CNN) based UNET for automatic detection and extraction of geophysical anomaly features at a wide range of scales using convolution filters. The input images are passed through a series of downsampling filters to retain their bare essential features. These are then upsampled to resemble the desired geophysical anomaly patterns persisting at multiple depth levels, following which they are extracted and examined in the geological context. The workflow is demonstrated for mapping major structural features associated with carbonatite-alkaline complexes in a study area in NW India. Manually derived and mapped structures were used to train the algorithm. As such algorithms require a large amount of training data, the input and training data were segmented to generate a robust dataset. The algorithm successfully delineated major structural and morphological features that conform with the visually mapped and interpreted features. An intersection over union (IOU) ratio of ~0.8 was obtained for the test images. This workflow can potentially improve the efficiency of prospectivity assessments of critical element deposits associated with carbonatite-alkaline complexes and can be applied to discover carbonatite-alkaline ring complexes from geophysical data in deeply weathered and under-cover terrains.

Long-term Relief Evolution of the Andean Chain in the Bongará Region (Northern Peru): Implications for the Genesis of Supergene Ore Deposits

Anna Sorrentino[1], Nicola Mondillo[1, 2], Ettore Valente[1]
1. University of Naples Federico II, Naples, Italy, 2. Natural History Museum, London, United Kingdom

A morphometric approach, based on the investigation of the topography and river network features of two hydrographic basins, Utcubamba and Chiriaco, was applied to determine the relationship between the irregular distribution of vertical motions (e.g., surface uplift) and the genesis of supergene Zn deposits located in the Bongará district, in the north-eastern sector of the Peruvian Andes. In the area three mixed supergene-hypogene ore deposits have been recognized: Mina Grande, Cristal and Florida Canyon. The Florida Canyon deposit falls within the Eastern Cordillera, in the Utcubamba Basin, and represents one of the most important Zn-Pb MVT deposits hosted in the carbonate and former evaporite-bearing rocks of the Pucarà Group. The mineralization consists mainly of sulfides and only one-third of the resource is nonsulfides and is hosted in Triassic rocks. The Mina Grande and Cristal deposits fall within the western margin of the Subandean Fold-and-Thrust Belt, in the Chiriaco Basin, and consist mainly of nonsulfide ores with less sulfides hosted in more recent Jurassic rocks.

The research has been carried out through the evaluation of the parameters such as elevation, local relief, swath profile, river longitudinal profiles, slope/area analysis to derive the normalized channel steepness index (*ksn*) and transformed river profiles (χ-long profiles). The spatial distribution of such indexes allowed to derive some consideration about the spatial distribution of vertical motions (e.g., surface uplift). Furthermore, the sharp increase in maximum, mean and minimum elevation moving from the Utcubamba Basin to the Chiriaco Basin, coupled with a jump in the mean Ksn values, suggest that the Chiriaco Basin has experienced either more recent surface uplift or surface uplift at rates higher than the Utcubamba Basin. This trend seems to be confirmed by the uneven erosion that the Floridia Canyon and the Mina Grande – Cristal deposits have experienced.

Improved Measurement for Processing of Critical Minerals

Henry Kurth
Scantech International Pty Ltd, Camden Park, SA, Australia

Application of innovative measurement technologies maximises the processing and recovery of critical resources. Improvements in process control require appropriate measurements that are representative, real time and precise. High specification prompt gamma neutron activation analysis (PGNAA) for elemental analysis and microwave transmission for moisture analysis have been successfully applied in many key commodities because they fully penetrate conveyed flows and measure continuously irrespective of particle size, belt speed, dust, and layering and segregation within the flows. Applications in iron ore, manganese, copper, cobalt, nickel, lead, zinc, bauxite, gold, silver, PGMs and lithium have improved processing through coarse waste rejection, blending, feed forward control, and monitoring quality of ores and products. The same technologies are used in recycling and are particularly suited to scrap metals, e-waste and black mass. Analysis of these materials enables batches of similar materials to be more efficiently processed and minimises product contamination. Highly customisable configurations provide high precisions over short analysis times and therefore fast responses. Measurement of moisture allows dry tonnage determination, better control over dust, and identifying potential handling issues where material is sticky or clumps easily due to clay or organics content. Efficient processing and quality management reduces the resources needed to process the materials and generates greater recoveries at lower cost. The paper provides examples of the applications in critical minerals and examines some benefits through the use of geometallurgy and sensor fusion in predicting process impacts as feed quality changes.

Optimization of Mineral Exploration Borehole Planning Using Partially Observable Markov Decision Processes with Belief-based Rewards

Tyler R. Hall[1], David Z. Yin[1], Robert J. Moss[2], Jef Caers[1]
1. Department of Geological Sciences, Stanford University, Stanford, CA, USA, 2. Department of Computer Science, Stanford University, Stanford, CA, USA

The placement, orientation, depth, and sequence of exploration boreholes controls rate of discovery and orebody delineation. The current state-of-the-art is to determine an optimal pattern of exploration boreholes based on expert judgement. However, using these purely expert-based methods does not incorporate quantitative optimization with stochastic geological models which more accurately reflect uncertainty of the subsurface. Here, we formulate mineral exploration as a partially observable Markov decision process (POMDP), and determine the optimal configuration and sequence of exploration boreholes by solving the POMDP using a reward function based on efficacy of information. Due to the complexity of the subsurface and many possible configurations of boreholes, the solution to optimal borehole placement is difficult to compute. In order to address this, we need to choose and apply approximations in exploration borehole optimization. Three approximations are studied: complexity of geological models, the detail of drilling observations, and the set of candidates of proposed boreholes. In our formulation for exploration drilling, geological features are parameterized as simple geometric objects (rather than wireframe or implicitly modeled objects), which enables quantified uncertainty reduction on economic parameters of an orebody such as size, shape, and depth, or combinations thereof (e.g. volume). We study the impact of using fully detailed versus simplified versions of borehole logs when updating the geological model with new data, and find that using fully detailed observations is not necessary. Finally, we optimize for sets of proposed boreholes, and also show optimal targeting when the locations of proposed boreholes are unknown. We demonstrate our results on a case study in the Midcontinent Rift.

Roadmap to Multi-scale 3D Geological Modelling from Drill-core Hyperspectral Data

Roberto De La Rosa, Moritz Kirsch, Samuel T. Thiele, Raimon Tolosana-Delgado, Richard Gloaguen
Helmholtz Institute Freiberg for Resource Technology, Freiberg, Germany

Drill-core extraction is essential for the characterization of deposits in mineral exploration, but the traditional logging process is fraught by biases and inconsistencies. Hyperspectral drill-core scanning can complement traditional logging techniques by providing continuous, high-resolution mineralogical data over the length of entire boreholes in a fast, reproducible, non-destructive, and cost-effective way. Hyperspectral drill-core scanning is by now widely applied in the mining industry. However, the results are big datasets that are challenging to interpret, validate and incorporate into the traditional workflows. In this contribution we showcase a procedure to increase the value of hyperspectral imaging to an exploration campaign by incorporating the following key elements: I) Automatic pre-processing of large, deposit-scale hyperspectral datasets. II) Unsupervised image processing to attain mineral and alteration maps as well as domains that are consistent across the entire dataset, as a basis for sample selection. III) Supervised classification of hyperspectral data for quantitative mineral abundance predictions along boreholes, using quantitative mineralogical data as training input and dictionary learning, an efficient machine learning technique. IV) Transformation of mineral abundance predictions into multi-scale and geologically meaningful lithological domains via a segmentation algorithm based on a continuous wavelet transform and subsequent classification. Finally, we present a multi-scale approach to correlate these derived domains for 3D geomodelling based on implicit scalar field representations, where the boundaries to be interpolated are identified as marker horizons and are further informed with mineralogical compositional data. In multi-scale modelling, the choice of the scale to work with is made by the user, where domains at larger scales are used for regional modelling, and smaller scale domains are used for modelling with higher resolution within the sub-blocks.

Predictive Mineralogy in Mineral Exploration and Waste Rock Management

Katharina Pfaff[1], Daniel Schmidt[1], Brian Goolsby[2], Chad Ostrander[2], Nigel M. Kelly[3], Jonathan Knapp[3], Thomas Monecke[1], Aaron Noble[4], Annelie Lundström[5]
1. Center for Advanced Subsurface Earth Resource Models, Colorado School of Mines, Golden, CO, USA, 2. Hitachi High-Tech Canada, Toronto, ON, Canada, 3. Bruker Nano Analytics, Madison, WI, USA, 4. Virginia Tech, Mining and Minerals Engineering, Blacksburg, VA, USA, 5. Minalyze AB, Sävedalen, Sweden

Quantitative scanning electron microscopy (automated mineralogy) has been extensively used in the mineral processing industry to characterize ore and tailings to evaluate the efficiency of mineral extraction. However, the data-rich product of automated mineralogical analysis suggests powerful application in a predictive context to solve a range of issues throughout the mine life cycle. The ongoing study described here demonstrates that automated mineralogical analysis on exploration drill core may be useful beyond developing exploration models but may also be used to predict the potential environmental impact of future waste rock.

In this study, exploration drill core from the Idaho cobalt belt was analyzed using continuous XRF core scanning, allowing classification of the wall-rocks surrounding the ore zones into different waste rock types through cluster analysis. These different waste rock types were subsampled for automated mineralogical analysis to fully characterize their mineralogy and textural characteristics to predict their acid generating potential. This project demonstrates that such a holistic approach to rock characterization using data generated during exploration yields significant value to mining companies. By properly applying the mineralogical data, companies can thoroughly assess the environmental impact of mining early in the development cycle, allowing them to optimize their strategies for mine waste management and ultimately reduce their environmental footprint. Effluent waters derived from waste material that has a low probability of generating acid would need little or no treatment prior to release into the environment whereas waters emitted from waste materials that are producing acids would need to be mitigated.

Proactive waste management strategies, such as waste sorting and reutilization could significantly reduce operation costs, including costs associated with environmental mediation. Future research will be conducted on exploitable mineral precipitates formed from acid mine drainage to investigate whether their chemical composition (including their critical minerals potential) can be predicted.

Prospective Mapping of Carbonatite-associated Iron Oxide Mineralization in the Western Oulad Dlim Massif: Remote Sensing, Field, and Geochemical Investigations

Cheikh Elwali Malainine[1], Otmane Raji[1], Muhammad Ouabid[1], Jean-Louis Bodinier[1, 2]
1. Mohammed VI Polytechnic University, Geology and Sustainable Mining Department, Benguerir, Morocco,
2. Université de Montpellier & CNRS, Geosciences Montpellier, Montpellier, France

Over the past decade, the growing need for critical metals for the energy transition has resulted in a sharp increase in exploration programs focused on carbonatite complexes. Carbonatite/alkaline intrusive complexes and their associated deposits are known to host concentrations of this type of elements, in particular rare earth elements (REE). Additionally, post-emplacement processes often lead to secondary enrichment of several mineral resources, including REE, niobium and phosphates. In this study, we investigate the metallogenic potential of the western Oulad Dlim massif, in the northwestern margin of the West African Craton (south Morocco), where new carbonatite complexes have been recognized. The adopted approach is based on the application of multispectral data for the remote detection of carbonatite-related occurrences potentially enriched in REE and Nb. The areas of interest are validated by field inspections using in-situ portable X-Ray fluorescence analysis. In addition, further laboratory analyses are carried out, involving petrographic analysis, major- and trace-element characterization of rocks and ore mineral phases. Results reveal that the fusion of processed data of ASTER and Sentienl-2A sensors is reliable to delineate iron rich-zones within the carbonatite complexes. Remote sensing results are consistent with field observations, where 90% of validated occurrences are concentrated in only 10% of the total investigated area. Additionally, petrographic, mineralogical, and geochemical characterization confirms the presence of significant concentrations of REE, Nb and phosphate (up to 3.6 wt.%, 1.8 wt.% and 9 wt.%, respectively). These concentrations are associated with the presence of monazite, pyrochlore and apatite mineralization concentrated in the mapped iron ore formations. Our study has highlighted the existence of several high potential zones with great similarities in terms of mineralization in several annular structures of the western Oulad Dlim massif. Further exploration should be pursued in this area as it is a promising district for critical metal deposits.

Using AI to Extract Information from Legacy Documents

Amit Juneja, Henri Blondelle, Oriane Nguyen-Thuyet
Agile Data Decisons, Houston, TX, USA

For ore deposit exploration and modeling, the data stored in legacy documents is a key piece of information to make decisions on basins, claims, or leases.

Up to now, extracting this information from documents has been a very labor-intensive task that has rarely been done exhaustively on large amounts of documents. As a result, some key decisions are not taken using all the available data.

The development of artificial intelligence, machine learning, and natural language processing has introduced some new possibilities to get value from legacy documents at scale. Using two case studies we have developed for two major mining companies, we will present different types of methods and algorithms used to capture information in documents like:

- The document classification according to the document text and graphical content
- The capture of collar data such as the Drill Hole name, the Coordinates, the TD, Dip, Bearing, etc.
- The capture of values in assay tables and certificates
- The capture of figures in documents, like maps, logs, cross-sections or diagrams
- The detection of specific geological intervals in composite logs
- The detection of positive recommendations or negative conclusions in reports

The presentation will focus on the role of subject matter experts while building information detection models with machine learning or deep learning tools. The methodologies to measure the accuracy of the models will be detailed with examples to expose what can be achieved with the current technology to get trustable and actionable data.

Integration of Magnetic and Mineral Occurrence Datasets Over Az, Co, Nm, and Ut: Implications for Archean-Cambrian Basement Control on Jurassic-Eocene Mineralization

Robert D. Charnock[1, 2], Ken Witherly[2], Thomas Monecke[1]
1. Center for Mineral Resources Science, Department of Geology and Geological Engineering, Colorado School of Mines, Golden, CO, USA, 2. Condor Consulting, Lakewood, CO, USA

With the demand for metals and critical elements increasing and the discovery of new deposits decreasing, it is increasingly important for regional exploration programs to reduce large and complex datasets into district scale exploration targets. We present an airborne magnetic dataset and mineral occurrence dataset covering AZ, CO, NM, and UT that has been processed to reduce complexity. Magnetic lineaments (high magnetic “ridges”, low magnetic “troughs”, and magnetic source edges) were semi-automatically generated to a depth of 20 km, in 1 km increments. Mineral occurrence data was parsed for Ag, Au, Cu, Mo, Pb, and Zn occurrences. The two datasets were compared to each other and to major geologic features, which tend to be crustal in extent and regional to continental in scale. Areas of moderate to high mineral occurrence density are almost always parallel along distinct and (semi)-continuous magnetic lineaments for upwards of hundreds of kilometers. These mineralized-associated lineaments are spatially related with E-trending Archean and Cretaceous to Eocene; NE-trending Paleoproterozoic; NW-trending Paleo- and Mesoproterozoic, Cambrian, and Pennsylvanian; and NNW- to NNE-trending Jurassic and Cretaceous to Miocene ancient continental margins, basement arc terranes, terrane boundaries, and overprinting orogenic-related structures. Areas of the highest mineral occurrence density are located along NE-, NW-, and E-trending lineaments, especially where they intersect. These relationships suggest that mineralization (dominantly Jurassic to Miocene in age in AZ, CO, NM, and UT) related to magmatism and metamorphism during the Nevadan-Sevier-Laramide orogenies and later extension, was strongly controlled at the regional scale by Archean to Cambrian basement-building features. Occasionally, the same Proterozoic features are associated with mineralization on either side of the unmineralized Colorado Plateau. Using these regional mineralized-associated lineaments and structures as guides, district-scale exploration targets can be defined.

Gold Mineralization at the Tuvatu Alkalic Epithermal Au-Ag Telluride Deposit, Viti Levu, Fiji

Jake A. Jefferson[1], Thomas Monecke[1], Sergio Cattalani[2]
1. Colorado School of Mines, Golden, CO, USA, 2. Lion One Limited, Nadi, Fiji

The Tuvatu deposit is an alkalic epithermal Au-Ag telluride deposit located on the island of Viti Levu in Fiji. The deposit lies in the southwestern corner of the Navilawa caldera along the ENE-trending Viti Levu Lineament, which controls the location of a series of important alkalic epithermal deposits. The Tuvatu deposit is hosted within a ~4.85 Ma monzonite that intruded into the Neogene volcanic succession exposed within the Navilawa caldera.

Gold mineralization at Tuvatu occurs over a strike length of almost 900 m, has a vertical extent of more than 500 m, and remains open at depth. Different vein stages have been recognized. Most of the gold occurs as native gold in paragenetically late quartz veins that are typically narrow (<4 cm). In addition to native gold, telluride minerals occur in variable amounts in these quartz veins, and include calaverite, sylvanite, hessite, stützite, altaite, and coloradoite. The native gold and telluride minerals form dendritic aggregates that occur together with delicate, fan-shaped roscoelite aggregates within a matrix of mosaic quartz. The mosaic quartz is interpreted to have formed though recrystallization of non-crystalline silica. As part of this study, pyrite contained in the quartz veins was investigated for the presence of invisible gold. Backscattered electron imaging of pyrite grains showed that As zoning is common. LA-ICP-MS trace element analysis confirmed that there is a close correlation between As and Au enrichment in pyrite and demonstrated that pyrite at Tuvatu contains invisible gold. Ongoing investigations focus on the quantitative evaluation of the contribution of invisible gold to the overall gold budget at Tuvatu. Although existing metallurgical studies indicate expected gold recovery of 87 to 92 percent at Tuvatu, invisible gold could have implications for future recovery at this and other alkalic epithermal Au-Ag telluride deposits on Viti Levu.

The Magmatic Duration of Climax-type Porphyry Mo Systems

Joshua M. Rosera[1], Sean P. Gaynor[2], Ryan E. Frazer[3]
1. United States Geological Survey, Energy & Minerals Science Center, Reston, VA, USA, 2. University of Geneva, Department of Earth Sciences, Geneva, Switzerland, 3. United States Geological Survey, Geosciences and Environmental Change Science Center, Denver, CO, USA

The duration of magmatism related to porphyry mineral systems is receiving increasing interest with respect to evaluating metal endowments and formation mechanisms. Most studies are focused on economically important porphyry Cu-Mo-Au systems, but porphyry Mo deposits are another significant global resource for base and critical mineral commodities. Although attempts have been made to relate magmatic duration to Mo endowment, the input geochronological data are mostly imprecise and/or inaccurate. For example, mineralization at the giant Climax Mo deposit in Colorado is often cited as occurring between 33–25 Ma. However, the oldest age is an average of zircon fission track dates that range from 50–17 Ma, and thus has a large uncertainty (33 ± 4.2 Ma, 2σ) (White and others, 1981). The Chalk Mountain topaz rhyolite that erupted from the Climax magma system initially had a reported K-Ar age with large uncertainty (27.7 ± 3.8 Ma), but new high-precision U-Pb zircon geochronology demonstrate it erupted at 26.299 ± 0.028 Ma. Molybdenum mineralizing intrusions elsewhere in Colorado have also yielded new higher-precision U-Pb zircon mean ages that differ by 1–2 Myr from previous dates. These differences are on the order of the total duration of mineralization for the largest porphyry Cu deposits (2–3 Myr), and are problematic for any attempt to model mineralization longevity. Instead, multi-sample high-precision geochronology studies are required. For example, high-precision geochronology indicates that the Questa porphyry Mo deposit (northern New Mexico) formed by multiple short-lived pulses over <500 kyr. Unmineralized fluorite and topaz-bearing leucogranites related to Oligocene extension in the Southern Rocky Mountains are also known to have been assembled in <1 Myr. More data are required, but we suggest that the total duration of episodic porphyry Mo mineralization is unlikely to exceed that of well-characterized porphyry Cu systems.

Geoenvironmental Properties of the Quebrada Blanca Deposit Waste Rock Using Hyperspectral Mineralogy

Enrique Saez Salgado[1], Anita Parbhakar-Fox[1], Regina Baumgartner[2], Nathan Fox[1]
1. The University of Queensland, Brisbane, QLD, Australia, 2. Teck Metals Ltd, Trail, BC, Canada

Chile produced 5.73 Mt of copper by 2020, primarily resulting from the exploitation of porphyry copper-type deposits. Considering cut-off grades averaging 0.5% it is estimated that 99% of the mined materials are non-ore materials (e.g., waste rock, spent heap leach, tailings, etc.). Chilean mine sites produce 530 Mt of tailings per year, with approximately 1.5 Mt deposited every day. Recent calculations confirm that ~11.0 kMt of tailings are distributed in 757 tailings storage facilities. Commonly, these non-ore materials can contain pyrite (FeS_2), which under surficial conditions, can oxidize to produce acid and metalliferous drainage (AMD). However, non-ore materials can also represent vital sources of supplementary metal commodities (primary and critical metals) that can be recovered after mineral reprocessing of tailings or directly from waste rock.

Traditionally, the environmental characteristics of future waste materials are defined using established chemical tests (e.g., static and kinetic tests). However, these do not routinely consider the measurement of mineralogical properties in detail. To address this gap, the study aims to develop a 'predictive environmental reactivity assessment toolkit'. The approach is being developed using drill core materials at the Quebrada Blanca Cu-Mo porphyry hypogene deposit located in Northern Chile to predict the properties of future non-ore materials. The research program will evaluate the mineralogy of samples across different lithologies and alteration zones, including environmental forecasting as well as potential source for supplemental extraction, using the combination of: (1) clustering analysis of available raw hyperspectral data; (2) industry-standard chemical tests; and (3) mineralogical and geochemical quantification. The expectation is to generate geoenvironmental domains based on hyperspectral mineralogy, environmental reactivity in a more selective way and assist with future non-ore materials management at early life-of-mine stages.

Alteration Targeting and Geophysical Exploration of the Little Hatchet Mountains, New Mexico

Kenneth Singh[1], Philip Goodell[1], Antonio Arribas[1], Michael Jacobs[2], Sergio Macias[1]
1. University of Texas at El Paso, El Paso, TX, USA, 2. Geosyntec Consultants, Seattle, WA, USA

As part of the southwestern North American porphyry copper province, mining of natural resources in New Mexico has historically played a crucial role in economic development for over 150 years and includes this project area in the Little Hatchet Mountains. The northern half of the Little Hatchet Mountains lies in Grant County and includes the Eureka Mining District (EMD) whereas the southern half is in Hidalgo County and includes the Sylvanite Mining District (SMD). The EMD, a copper and silver-lead mining district from 1880-1961, had a total estimated production from the Laramide veins of 2.9 million lbs. Pb, 1.7 million lbs. Zn, 500,000 lbs. Cu, 5,000 ounces Au, and 450,000 ounces Ag. The southern half is home to SMD, a silver-gold mining district from Laramide skarn, vein, and placer deposits occur in the district and production including 2,500 ounces Au, 130,000 lbs. Cu, and 8,000 lbs. Pb from 1902-1957. This study focuses on remotely sensed mineral alteration zones detected using Mars 2014 USGS shapefiles of hydrothermal alteration in the Basin and Range province, which allowed for the targeting of electromagnetic surveys that have been conducted, and electrical resistivity surveys to be conducted of potential mineral deposits. Electromagnetics survey readings, using the Geonics EM34-3XL that reaches a depth of ~180 ft, have resulted in areas of interest that are most likely mineralized. These surveys have also allowed us to determine where the electrical resistivity surveys, using the SuperSting R8 112 electrode 5 m interval spacing for a max depth profile of 555 m, will be performed. Spectral microscopy and x-ray diffraction on up to 18 samples was conducted to test remote sensing ground truth certainty. The present study is enhancing the knowledge base of these deposits and southwest New Mexico.

New U/Pb Geochronology from the Proterozoic Penokean Orogen, Wisconsin: Implications for VMS Metallogeny

Evan M. Weber[1], Robert W. Lodge[1], Jeffrey Marsh[2]
1. UW-Eau Claire, Eau Claire, WI, USA, 2. Laurentian University, Sudbury, ON, Canada

This study presents new U/Pb and trace element data from zircons obtained from volcanic and intrusive rocks from the Pembine-Wausau terrane of the Paleoproterozoic Penokean Orogen in northern Wisconsin. This terrane is best known for hosting almost 150 Mt of volcanogenic massive sulfide (VMS) mineralization. However, challenging socio-political obstacles preventing mining and exploration activities coupled with extensive overlying Paleozoic and Quaternary deposits has limited the study of the terrane. The current need for critical minerals highlights the importance of improving our understanding of this underexplored and poorly exposed orogen. This study aims to provide a better tectonostratigraphic understanding of the volcanic terranes to improve regional tectonic and metallogenic models for base and precious metals. Samples were collected from within the Ladysmith-Rhinelander and Eau Claire volcanic belts, and zircons from these samples were then analyzed at Laurentian University (Sudbury, Ontario, Canada) via Split-Stream Laser Ablation Inductively Coupled Plasma Mass Spectrometer (LASS-ICP-MS) to obtain U-Pb, Hf-Lu, and trace element data. Preliminary results reveal complex age relationships and basement architectures that were previously unrecognized. The samples obtained from the Eisenbrey deposit within the Ladysmith-Rhinelander belt temporally correlate with felsic rocks from the Back Forty deposit in Michigan revealing a previously unrecognized wide-spread younger VMS forming event. Samples from the Lynne deposit reveal the presence of a co-eval subvolcanic intrusion in the presence of Archean basement. Samples from the Eau Claire volcanic complex reveal ages that correlate with other VMS-forming strata across the Pembine-Wausau terrane. Therefore, tectonic models for the development of the Penokean Orogen and the formation of VMS deposits need to be revisited.

Carbonate-Apatite in the Southwestern Ordos Basin, China: Link with Life and Its Significance

Chao Zhang[1], Xiaobin Wu[1], Baojin Zhao[1, 2]
1. Southwest Petroleum University, Chengdu, Sichuan Province, China, 2. Department of Environmental Sciences, University of South Africa, Johannesburg, South Africa

The enrichment of life in ancient sedimentary rocks has been widely recognized, but the direct link between mineralization and organism remains unknown for many cases. The shale of the 7th member of the Triassic Yanchang Formation from the Ordos basin in China is selected for the study of apatite whose morphologies and chemical compositions were examined by SEM/EDS, together with XRD and other methods. The typical apatite assemblages comprise at least four types alike: 1) Cylinder; 2) Rod; 3) Nodule, with an outer shell, uncompleted inner kerogen shell, and cauliflower core; and 4) Teeth, while their overall sizes of >10 micrometer are much larger than microorganisms of 1 to 5 micometer. We concluded that 1) The carbonate apatite resemble the body parts of organisms; 2) The mineral chemistry is characterised by enrichment of carbon, so that it is bio carbonate apatite ($Ca_5(PO_4, CO_3)_3(F)$, which is significantly different from the igneous origin ($Ca_5(PO_4)(F, OH, Cl)$; 3) Kerogen in bio-carbonate apatite is typically derived from microorganisms, animal remains or plant sources; 4) Bio carbonate apatite is a typical amorphous form rather than crystals so that there is no hydrothermal activities involved in the formation; 5) Phosphate mineralisation by life in shales, specifically in the marine environment is an important type of mineralisation, which is useful in future exploration of phosphate deposits in the similar environment; 6) Existence and enrichment of apatite in oil shale the Ordos bains or others might be useful for agricultural cultivation; and 7) Apatite in the life cycling of phosphate in the Ordos basin can be further analysed for its oxygen isotopes as original temperature proxy and REE for determination of the sedimentary environmental in the near future.

Framboidal Pyrite Formation and Its Significance in the Southwestern Ordos Basin, China

Chao Zhang[1], Xiaobin Wu[1], Baojin Zhao[1, 2, 3]
1. School of Geoscience and Technology, Southwest Petroleum University, Chengdu, China, 2. Department of Environmental Sciences, University of South Africa, Florida, South Africa, 3. Overseas Expertise Center for Deep Marine Shale Gas Efficient Development Innovation (111Center), Southwest Petroleum University, Chengdu, China

Enrichment of framboidal pyrite was identified in the Yanchang Formation's Section 7 (Chang 7) in the Triassic of the southwestern Ordos Basin, China, which is the second largest in oil and gas reserve and production. The basin is located in the south margin of the North China Craton and bounded by a number of faulted blocks and fold belts in its margins, which basement is a metamorphic system of the Archaean and the lower Proterozoic, while the basin itself consists of from the Precambrian to Tertiary systems, with a total thickness of 5,000 to 10,000 m and 370,000 km^2 crossing Shaanxi, Gansu and Shanxi Provinces. This study was focused on the bottom portion of Chang 7, being characterized by both higher Gamma Ray American Petroleum Institute (GR API) and Self Potential milli Volt (SP mV). The methods used include SEM/EDS, together with XRD and ICP-OES etc. The results show that most of the pyrite aggregates are typically spherical framboids with almost unique numeber of individual cuboctahedron crystals of pyrite with almost perfect crystography whether or not bigger or samll in size. The framboidal pyrite aggregates are largely enlogated and worm- or green bean- like shapes except for minor individual pyrite framboids, largely along the bedding of the sediments. There is no clear link with hydrothermal and tectonic activities, and directional growth feature. A certain membrane structure can be seen on the outer shells and nets of pyrite framboids, indicated by predominant organic carbon, ditected by SEM/EDS. We conclude that enrichmentment of framboidal pyrite is strongly connected with life, specifically microorganisms and organic matters, which is a typical indicator of the oil source rocks. Enrichment of prite may generate acid mine drainage and cause environmental damage in exploitation of oil, gas and shale gas.

What Governed Au-Cu Mineralization Fertility of Magmatic Systems?: an Example from Yinan Au-Cu Skarn Deposits in Luxi District, North China Craton

Si-Yuan Li[1, 2], Si-Shen Li[3, 4], Zhuang Duan[5], Jian-Wei Li[1, 2]
1. School of Earth Resources, China University of Geosciences, Wuhan, China, 2. State Key Laboratory of Geological Processes and Mineral Resources, China University of Geosciences, Wuhan, China, 3. College of Automobile Engineering, Jilin University, Changchun, China, 4. State Key Laboratory of Automobile Simulation and Control, Jilin University, Changchun, China, 5. Institute of Geophysical and Geochemical Exploration, Chinese Academy of Geological Sciences, Langfang, China

Au-Cu ore-forming magmas are commonly thought to have been enriched in metal and/or volatile. An intensive study of whole rock samples, zircon, and apatite from magmatic systems associated with Yinan Au-Cu skarn deposits was undertaken to test these assumptions. Combined, in-situ Hf isotope analyses in zircon and bulk rock composition and Sr-Nd isotope imply that the Yinan ore-related intrusions were derived from an enriched lithospheric mantle metasomatized by subducted derived fluids and/or melts. The geochemical signatures of whole rock samples, zircon, and apatite unanimously indicate ore-related intrusions are characterized by high oxygen fugacity with volatile (H_2O, S, and Cl) enriched, most likely from the source inheritance. Petrographic, petrological and geochemical evidence reveal substantial sulphides were coprecipitated with pyroxene and amphibole during the magma evolution and trapped in cumulates, which may resulting in much lower Au and Cu content of magmas. These results implies that the formation of metal-rich cumulates at depth was not detrimental to their fertility, although likely loss of metal. Observed magmatic sulphides are tiny and are mostly found in non-fractured silicate minerals, making them inefficient for accumulating Au and Cu and difficult to be subsequently remobilized. High oxygen fugacity and high Sr/Y values of Yinan Au-Cu associated intrusions could not be produced by partial melting of those sulfide-bearing cumulates. These facts imply limitted contribution of sulfide-bearing cumulates to later Au-Cu mineralization. Given that such Au, Cu-depleted melts were nonetheless able to generate Au-Cu deposits, our results highlight magma volatile concentrations, not Au or Cu concentration, is the most important factor in governing Au-Cu mineralization potential.

Combined pXRF and FTIR assays for Exploration and Grade control; Examples from SGS Field Analytical Services and Testing (FAST) Toolbox

Hugh DeSouza[2], Alexander Seyfarth[1], John Woods[3]
1. SGS Natural Resources, JANESVILLE, WI, USA, 2. SGS Natural Resources, Toronto, ON, Canada, 3. SGS Natural Resources, Perth, WA, Australia

The talk focuses on the use of portable XRF(pXRF) and FT-IR spectroscopy on analytical grade pulp for exploration, grade and mill control. By using proven setups for sample preparation an analytical grade pulp (85% passing 75 micron) is analyzed with FT-IR using an Attenuated Total Reflection (ATR) setup as well as a loose powder with a modern pXRF. The talk will cover the limitations of each technology and the calibration approaches used by SGS to ensure fit for purpose data. We will show application examples for critical metal deposits such as Pegmatitic Lithium as well as for Bauxite deposits. Benefits of such combined chemical and mineralogical dataset's for modelling and use of calculated mineralogy are discussed for each of the two application scenarios.

XRF Core Scanning; Continuous Scanning Versus Point Scan: a Practical Review

Alexander Seyfarth[1], Thomas Monecke[2]
1. SGS Natural Resources, JANESVILLE, WI, USA, 2. Colorado School of Mines, Golden, CO, USA

Chemical assays, directly on fresh or historical core, with modern Energy Dispersive X-Ray spectroscopy (EDX) are not longer an academic exercise but commercially available solutions. Several different service providers and instruments are in the market offering various solutions and setups. In the talk we are comparing the performance of a continuous EDX core scanning system (Minalyze CS) on core and RC Chips with an automated point scan system using a modern pXRF, implemented as the Geochemical Research Assistant (GERDA). We will cover the method limitations with respect to the origination of the X-ray signal, e.g. analyzed layer and the instrument calibration(s).

Instead of a head to head detection and performance approach of the vendor specs we are presenting the respective ideal and non-ideal use cases with different examples from the collection of core and chips by Colorado School of Mines. Typical machine throughput experienced in operation as well as well as additional deliverables will be compared as well.

SEG 2022 Conference: Minerals For Our Future

Unraveling the Geological History of the Fenelon Gold Deposit, Québec

Joy M. Carter, Daniel Gregory
University of Toronto, Toronto, ON, Canada

The Fenelon Gold Deposit is located in west-central Quebec, within the northern Abitibi Greenstone Belt. It is situated along the northern margin of the Sunday Lake Deformation Zone, a major east-west structural corridor that also hosts the Detour Lake Gold Mine. Until recent years, there has been little research and exploration conducted along the Sunday Lake Deformation Zone in comparison to its southern counterparts, the Cadillac-Larder and Porcupine-Destor Fault Zones. The Wallbridge Mining Company Ltd. acquired the Fenelon Property in 2018 and has since delineated significant gold systems along the margin of the Jeremie Diorite and hosted within crosscutting veins. The initial resource estimate of the Fenelon Gold Deposit was announced in November 2021, yielding 3.58 Moz gold in the indicated and inferred categories. This research project, one of the first undertaken at Fenelon, utilizes a combination of geochronology, whole-rock geochemistry, and sulfide mineral chemistry to characterize and unravel the timing of the formation of the host rocks and mineralization at the Fenelon Deposit.

This presentation discusses the results from TIMS and LA-ICP-MS analyses of zircons from the Jeremie Diorite and the surrounding sedimentary rocks, as well as Re-Os dating of a gold-bearing vein. These results are combined with observations from drill core to establish the chronology, and aid in the modeling of the deposit. Seequent's Leapfrog Geo software was used to model a geochemically distinct mafic phase of the Jeremie Diorite which may constrain mineralization toward the northeast. Lastly, LA-ICP-MS analyses of pyrite from the mineralized zones establish a relationship between sulfide and gold precipitation. These results enhance the geological understanding of the Fenelon deposit and will significantly contribute to resource expansion and exploration in the surrounding belt.

The Zaozigou Orogenic Gold-Antimony Deposit, West Qinling Orogen, China: Multiple Mineralization During Multiple Tectonic Evolutions

Kunfeng Qiu
China University of Geosciences, Beijing, Beijing, China

The hypothesis of multiple orogenic gold events within a single deposit remains controversial. The giant Zaozigou orogenic Au-Sb deposit in the West Qinling Orogen, central China, hosts two distinct mineralization styles: (1) steeply dipping orebodies with refractory Au in arsenopyrite and pyrite and (2) shallowly dipping orebodies with visible Au associated with stibnite. The framework for three major events in the Zaozigou deposit and their linkage with the multiple tectonic evolutions of the West Qinling Orogen is built based on field investigations and absolute ages. The deposit is hosted within the earliest Triassic folded sedimentary rocks, which are intruded by the Early to Middle Triassic ENE and WNW-striking dacite dikes. These pre-mineralization events are interpreted to indicate an E-W compression related to the subduction of the Paleo-Tethys oceanic slab (D1). The Au-Sb mineralization is divided into two phases. The Late Triassic mineralization (D2) formed along ENE-striking and steeply dipping sinistral faults. The shear zones are characterized by brittle to ductile deformation features, indicating a NNE-SSW compression. Monazite U-Pb geochronology return a date of 211 Ma, which is interpreted as the mineralization age and coincides with the Late Triassic collision between South China Block and South Qinling Terrane. The Early Cretaceous mineralization (D3) formed along shallowly dipping brittle fault zones, which indicate the local stress field as NE-SW compression. Dolomite U-Pb geochronology return a date of 137 Ma, interpreting this mineralization event as related to the Late Jurassic to Early Cretaceous reactivated collision between the South China and North China blocks. Therefore, the two distinct mineralization styles represent two discrete mineralization phases, with each different style of mineralization representing a separate period of fluid infiltration. The above evidence indicates that Zaozigou is the example to be recognized of a telescoped orogenic Au-Sb system.

Temporal Evolution of Copper Isotope in Dexing Porphyry Cu-Mo-Au Deposit, China, and Its Implications for Explorations and Cu Isotope Fractionation in Ore-forming Processes

Jiafeng He, Guoguang Wang*, Weiqiang Li, Pei Ni
State Key Laboratory for Mineral Deposits Research, Institute of Geo-Fluids, Frontiers Science Center for Critical Earth Material Cycling, School of Earth Sciences and Engineering, Nanjing University, Nanjing, China

Cu isotope analyses have provided direct constraints of supergene enrichment and ore-forming processes in porphyry copper deposits. While the heavy Cu-isotope signature of the enrichment zone is successfully illustrated by preferential partition of ^{65}Cu (the heavy copper isotope) into oxidized fluids in the leaching process, the understanding of Cu-isotope patterns in the hypogene zone is hindered by relatively poorly constrained Cu isotope fractionation behaviours in the ore-forming processes at high temperatures. We analyzed Cu isotope ratios of chalcopyrite from all four-stage veins in the Dexing giant porphyry Cu-Mo-Au deposit, China, and found a different Cu-isotope variation trend from the reported patterns. The Dexing-pattern features an increase of $\delta^{65}Cu$ (in permil) values from the innermost A vein samples (–0.26 ± 0.03‰, n = 3, 2SD) outward toward B and C vein samples (–0.12 ± 0.14‰, n = 19, 2SD), and a sudden drop to the outmost D vein samples (–0.42 ± 0.17‰, n = 9, 2SD). Combining the Cu-isotope measurements with previously published fluid inclusion data, we dipicted the detailed ore-forming processes of precipitation and multiple fluid boiling events by the Cu-isotope pattern. The enrichment of the heavy Cu isotope in B and C vein (the main stage of mineralization at Dexing) is similar to Northparkes porphyry Cu-Au deposit, Australia, where $\delta^{65}Cu$ values of sulfides drop from the high-grade core to the margin of orebodies. Deposit-scale Cu isotope studies suggest that the heavy copper isotope enrichment coincides with the high-grade ores in hypogene zones, which can be a vectoring tool for explorations. In addition, the overall low Cu-isotope signature (median = –0.17‰, n=31), relative to the average Cu isotope composition of bulk silicate Earth (0.06 ± 0.20‰, 2SD), may be caused by the tendency for ^{63}Cu (the light copper isotope) to partition into the exsolved fluid from silicate magmas.

Geochemical Data: a Critical Element in Geometallurgical Studies

Pim van Geffen, Neal Reynolds
CSA Global, Vancouver, BC, Canada

Geometallurgy, also known as geomet or orebody knowledge, is a cross-disciplinary field of geoscience and engineering that has become established over the past decade. Historically, the fields of mineral exploration, mining, and processing would be generally understood by geologists, mining engineers, and metallurgists, and knowledge would be shared among small teams involved in resource development and production. More recently, science and engineering have become more and more specialized and diversified, leading to siloing in universities and industry, and a dissociation of exploration geosciences from mining operations. This knowledge gap has grown to the point that a bridging discipline is now required to reconnect the disparate functions through designated teams of subject-matter experts using a common language. Mineral processing is primarily affected by mineralogy, which dictates grade, hardness, material handling, metal deportment, and recovery parameters. However, detailed mineralogical data are usually only acquired in small volumes of metallurgical or environmental testwork, or in the very early stages of exploration. High quality multi-element geochemical data are more commonly available for thousands of drill-hole intervals, providing the closest approximation of an orebody's mineral composition. In combination with appropriate metallurgical and environmental testwork, such geochemical data can play a critical role in the development of geometallurgical domains for resource models and in generating proxies for key indicators of ore-processing performance, tailings composition, and waste characterization.

Development of Data Systems to Support Critical Mineral Research in New Mexico

Virginia T. McLemore, Mark Leo-Russell, Amy Trivitt, Brandon Dennis, Dave Kasefang
New Mexico Bureau of Geology and Mineral Resources, New Mexico Institute of Mining and Technology, Socorro, NM, USA

The NM Bureau of Geology and Mineral Resources (NMBGMR) has accumulated data on mines and minerals resources, including critical minerals, in NM since 1927. Over past decades the NMBGMR has converted historical data into various electronic formats. Relational databases and images are available on the NMBGMR website with more content being added. To date, over 6800 photographs, 750 maps, and other historic documents have been uploaded onto the NMBGMR Photo Archive site. Under the USGS National Geological and Geophysical Data Preservation Program, USGS Earth Mapping Resources Initiative, and DOE CORE-CM programs, additional data including physical samples are collected and archived. Legacy chemical analyses are important to understand where critical minerals are likely to be found and their potential for development. More than 2900 chemical analyses from past projects are archived (https://geoinfo.nmt.edu/staff/mclemore/projects/mining/REE/McLemoreMasterChem_v5.xlsx). Not only do the actual chemistry data need to be accessible, but the context of such data is vital to our research. The context information includes locations, geological settings, collection methodologies, and in-house tracking of samples. Therefore, the NMBGMR is undertaking efforts to better organize, store, catalog, and retrieve a large collection of geochemical and related data. Previously, data were collected and analyzed in ad-hoc databases and spreadsheets. New SQL Server databases and web-based applications are being developed to enter, store, retrieve, analyze, and visualize these data on mines, mineral deposits, and samples. The purpose of these databases are to provide information that will aid in 1) evaluating critical minerals and other resource potential, 2) identifying their development and production, and 3) identifying possible environmental concerns. These data are needed by government agencies to make land-use decisions. In this presentation members of the NMBGMR economic geology and information technology teams will present the status of our efforts in developing these data systems, and lessons learned along the way.

Timing of Orogenic Gold-forming Events Related to the Tectonic Evolution of California

Ryan Taylor, Leah Morgan, Erin Marsh
USGS, Denver, CO, USA

Orogenic gold mineralization within the Sierra Nevada foothills (SNF) and Klamath Mountains (KM) provinces in central and northern California, occurred episodically during ~50 million years of the Late Jurassic through Early Cretaceous and is linked to major tectonic events. Provincial gold mineralization in the KM lasted from ~160–140 Ma, whereas gold mineralization in the SNF lasted from ~160–115 Ma.

Initial gold mineralization impacted the then-contiguous KM-SNF arc at ~160 Ma, coincident with major far-field plate reorganization in the Pacific basin and compressional conditions. The next period of orogenic gold formation occurred at 155-140 Ma, synchronous with sinistral transpression along crustal-scale faults. Beginning around 140 Ma, the KM area was offset westward from SNF and the active arc, leading to the cessation of magmatism and hydrothermal activity within the KM; however, both continued within the SNF.

Sinistral movement along major faults in the SNF persisted until ~125 Ma, when another far-field plate reorganization in the Pacific basin led to a reversal of strike-slip motion along these faults. Most gold deposits of the ~190 km-long Mother Lode belt in the SNF cluster between ~130-125 Ma and are interpreted to be related to this transition to dextral motion along the terrane-bounding faults. A few younger ages suggest another period of gold mineralization at ~115 Ma in the northernmost (Alleghany district) and southernmost (Bagby district) parts of the SNF gold province that correlates to changes in the direction of movement for the Pacific and Farallon plates.

Despite a general overlap in timing of regional post-accretion magmatism and orogenic gold mineralization, there is no genetic link between these two processes. Major age differences between magmatism and mineralization on the local scale are the norm. When magmatism and gold mineralization are temporally correlated, geochemistry of trace minerals indicate that they are not genetically related.

Importance of Foundational Petrology and Geochemistry for Evaluating Igneous Sources to Base Metal Mineralization in SW Ireland

Paul Slezak[1], Murray W. Hitzman[1], David van Acken[2], Eoin Dunlevy[1], David Chew[3], Foteini Drakou[3], Mark Holdstock[4]
1. Irish Centre for Research in Applied Geoscience, University College Dublin, Dublin, Ireland, 2. National Centre for Isotope Geology, University College Dublin, Dublin, Ireland, 3. Department of Geology, Trinity College Dublin, Dublin, Ireland, 4. Group Eleven Mining & Exploration Ltd, Dublin, Ireland

County Limerick contains the largest exposure of Carboniferous igneous rocks in Ireland, which occur as diatremes, lava flows, dykes, hypabyssal intrusions, and tuffs that are collectively known as the Limerick Igneous Suite (LIS). The LIS is comprised of two smaller, separate igneous packages: 1) the Knockroe igneous units, which range from alkaline basalts to trachyandesites, and 2) the Knockseefin igneous units, which range from alkaline basalts to basanites. Both units have undergone significant amounts of chlorite-calcite ±prehnite alteration, minor pyrite mineralization, and are spatially associated with multiple Zn-Pb sulfide prospects in SW Ireland.

Uranium–Pb dating of apatite establishes an igneous crystallisation age of c. 350 Ma for the Knockroe units. Bulk rock Sr and Nd isotopes from the least altered Knockroe samples range from 0.70301–0.70454 and 0.512457–0.512493, respectively. Strontium isotopes for the least altered Knockseefin samples range from 0.70325–0.70386 and the Nd values range from 0.512431–0.512437. Altered samples are buffered against changes in Nd, but some analyses show excursions towards radiogenic Sr, indicating contamination from Carboniferous seawater and/or introduction of Rb. The latter alteration returned ages within uncertainty of the U–Pb dates.

The world-class Lisheen and Silvermines deposits are ~50 km north of the LIS and have Re–Os ages of 346 ±3 Ma and 334 ±6 Ma, respectively, which correspond to the ages of the LIS. However, the Nd and Sr isotope signatures from the LIS are very different than those previously obtained from gangue carbonate and barite associated with hydrothermal Zn-Pb-Fe sulfides at the Navan Mine. Using the results from Navan as a proxy for Irish-type mineralization suggests the LIS is not necessarily a mineralization source. However, the U–Pb ages indicate the LIS may have been a heat source along the Iapetus Suture Zone, which could have helped to drive Irish-style mineralizing systems.

Sulfur Isotope Systematics of Granitoids from the Yilgarn Craton Shed New Light on the Fluid Reservoirs of Neoarchean Orogenic Gold Deposits

Stefano Caruso[1, 2], Marco L. Fiorentini[2], David C. Champion[3], Yongjun Lu[4, 2], Yuichiro Ueno[5], Robert H. Smithies[4, 6]
1. CSIRO, Kensington, WA, Australia, 2. The University of Western Australia, Crawley, WA, Australia, 3. Geoscience Australia, Canberra, ACT, Australia, 4. Geological Survey of Western Australia, Perth, WA, Australia, 5. Tokyo Institute of Technology, Tokyo, Japan, 6. Curtin University, Perth, WA, Australia

The identification of the fluid reservoirs of Neoarchean orogenic gold deposits is crucial for understanding the sulfur cycle through the lithosphere and improving exploration strategies to spark new discoveries. Recently, the analysis of Mass-Independent Fractionated Sulfur (MIF-S) isotope signatures of sulfide from gold-bearing assemblages has revealed that (i) at least some of the sulfur in the mineralizing fluids had previously interacted with the Archean oxygen-free atmosphere, and (ii) such fluids were commonly sourced from homogeneous reservoirs at depth. However, the nature and spatial distribution of such reservoirs has so far remained elusive as the lithological and compositional variation of the deep Archean lithosphere remains poorly constrained.

In this study, we focused on the sulfur isotope composition of granitoids, which can represent a proxy for constraining the signature of lower crustal sulfur reservoirs. We analyzed granitoids distributed throughout the Archean Yilgarn Craton of Western Australia, demonstrating that most of them carry MIF-S signatures. In view of recent studies on the genesis of granitoids advocating for magma formation through infra-crustal melting at mid- to lower-crustal regions, MIF-S signatures in granitoids indicate the presence of supracrustal sulfur transported to deep crustal levels and incorporated into the bulk melt sources. Furthermore, the spatial distribution of MIF-S signatures in the granitoids mimics that of Nd and Hf radiogenic isotopes and identifies homogeneous domains that coincide with the current terrane subdivision of the Yilgarn Craton, which reflects cratonic crustal evolution and lithospheric architecture. Because this spatial distribution in MIF-S signatures is also reflected in Neoarchean orogenic gold deposits, it implies a common sulfur reservoir. Therefore, we suggest that the mineralizing fluids exsolved from melts originated through the reworking of deep crustal regions, possibly with the contribution of portions of metasomatized mantle lithosphere.

Machine Learning for Lithologic and Ground Type Modeling at the Stillwater Mine

Gretchen E. Moore[2], Maureen Moore - Roth[1]
1. Maptek, Evergreen, CO, USA, 2. Sibanye Stillwater, Nye, MT, USA

Orebody knowledge and characterization are the foundation of economic decision-making from exploration to mine development. Timely and robust modeling of narrow vein-style deposits can be problematic in the fast-paced underground production environment. This presentation will highlight new methods to model the lithology and ground types in and near the narrow vein platinum and palladium deposit of the Stillwater Complex that is being mined in Nye, Montana, USA. The geology and geotechnical staff at the Stillwater Mine was challenged with modeling thin problematic poor ground areas that are prone to unraveling, and could be areas that are the root cause of falls of ground when encountered unexpectedly. In the case study, presenters will showcase how machine learning techniques coupled with conventional estimation methodologies were used to model the lithology and Q data. Utilizing a processing partnership between Maptek and Sibanye Stillwater to test and implement DomainMCF, a machine learning technique was leveraged to generate geologic domain boundaries for lithology and Q data. This approach to dynamic modeling has allowed the Stillwater Mine teams to rapidly evaluate new diamond drilling information. The lithologic model data is now used by the grade control geologists prior to mining new floors, guiding where they should be mining. This predictive method helps the geology-driven mining operation empower the forty-plus geologists and five plus rock mechanics engineers to guide the mining operation. Future uses of this modeling methodology are applications to digital face mapping for both teams. Additionally, testing of the DomainMCFTM modeling methodology as applied to the development of a structural model. Recent innovations and enhancements in machine learning for domain modeling allow for geologists to spend more time boots-on-the-ground with their geologic data to help more efficiently predict and use the geology and geotechnical information for efficient and economic mine development.

The Importance of Coherent Geologic Maps in Mineral Exploration

Dean M. Peterson, Gabriel J. Sweet
Big Rock Exploration, Duluth, MN, USA

The basis for most types of geologic investigation is fundamentally rooted in geologists' observations and interpretations made of landscapes, exposed rocks, and surficial materials in their natural habitat, "in the field". Coherent geologic maps, which take many years to create, represent assembled collections of observations in context of space and geologic time by teams of geologists. Such geologic maps are the foundation upon which mineral exploration programs and mineral resource developments are built. In today's mineral industry, geologic maps are largely digital compilations of publicly available regional/district scale GIS data (downloaded from geological survey websites) merged/overlain with detailed industry geologic mapping of prospects and/or project areas. For the most part, the mineral industry quickly compiles digital data into geologic maps and is seemingly always searching for new ways to quickly capture data in the field digitally. The ease in which the mineral industry can generate geologic map products today can be good, bad, or ugly.

The true strength of mineral exploration programs depends primarily on decisions based upon the integration of all pertinent information, large parts of which is housed in coherent digital geologic maps and the unseen GIS attribute table data behind these maps. Mineral deposit models have arisen as mental paradigms that act as guidebooks to help geologists process information, and effective development of models that enhance success rates and decrease costs is critically important to the mining industry as it explores new terrain and becomes constrained by stringent economic, environmental, and political bounds.

This talk will touch upon the creation of industry-based coherent digital geologic maps for use in mineral exploration programs with examples of how such maps can be used in targeting mineral exploration programs of orogenic gold, mafic magmatic, and volcanogenic massive sulfide mineral systems.

Newmont Geochemistry, the Untold Story

Simon A. Marshall[1], Jeffrey S. Bigelow[1], Kiel A. Arndt[2], Sarah Rice[2]
1. Newmont, Welshpool, WA, Australia, 2. Newmont, Denver, CO, USA

As Newmont moves into its second century, there comes an opportunity to reflect on a history of commitment to exploration, innovation and technology. Newmont's 100-year journey has populated the gold explorers' toolbox with geophysical techniques as notable as induced polarisation; however, the parallel story of the development of industry-leading geochemistry is yet untold. Although there are no silver bullets in gold exploration, a 40-year effort by Newmont's geochemistry team with a focus on sampling theory, data integration, and aggressive field programs has resulted in proprietary geochemical exploration tools which are still seeing active development and deployment.

Newmont's Bulk Leach Extractable Gold (BLEG) stream sediment sampling technique was developed over the 1980s and 1990s as an efficient and reliable low-level regional reconnaissance screening tool. As gold is a notoriously difficult element to analyse in stream sediments due to highly variable field duplicates, the technique was engineered to achieve acceptable reproducibility of results at the lowest detection limits available at the time. Amongst several discoveries, Newmont BLEG is credited for revealing the world class Batu Hijau Cu-Au porphyry system in Indonesia (Maula and Levet, 1996).

More recently, exploration has trended away from a focus on outcropping systems into the cryptic world of undercover targeting. This focus has seen Newmont's next generation of geochemical tools emerge; Deep Sensing Geochemistry (DSG) and Terrain Mapping Geochemistry (TMG). Although DSG was first named publicly in 2013, these techniques were developed following years of internal and external research in fine fraction geochemistry, applied field geochemical and regolith control, and advances in analytical laboratory capabilities using ultrapure digestions and instrumentation pushing elemental detection limits below crustal abundance.

Explorers today are experiencing a geochemical renaissance where these tools, leveraging applied geochemistry fundamentals are integrated with other advanced exploration tools and datasets to deliver the next generation of discoveries.

Are Non Invasive Methods of Exploration Efficient?

Richard Gloaguen, Sandra Lorenz, René Booysen, Moritz Kirsch, Margret Fuchs, Pedram Ghamisi, Sam Thiele, Yuleika Madriz
Helmholtz Institute Freiberg for Resource Technology, Freiberg, Germany

It is indubitable that the discovery of new deposits will be required to support our efforts towards energy transition and climate neutrality. At the same time, the growing opposition to mining threatens many new exploration projects. In fact, social attitude towards mining activities is becoming the main barrier to metal sourcing. New approaches to exploration are required and the use of remote techniques will be necessary to ensure the social acceptance of new projects. Remote sensing and airborne geophysics are good candidates but doubts remain regarding their ability to efficiently contribute to discovery. We recently set up three reference sites in Europe within the EU funded project INFACT. One of the objectives of our research was to benchmark new non-invasive exploration techniques not only in terms of technical efficiency but also their environmental impact and social acceptability. We assessed several geophysical methods such as ambient seismic noise airborne electromagnetics and drone-borne magnetics. We also conducted several experiments with hyperspectral imaging methods (VNIR-SWIR-LWIR) placed on a variety of platforms such as tripods, drones or airplanes. We developed an innovative 3D approach to process and display remote sensing data. We demonstrated that innovative geophysical and remote sensing technologies can contribute to discovery if they are used properly and provided guidelines. More importantly we showed that the integration of several technologies using machine learning techniques boost their potential. The 3D data help to characterize potential mineral deposits. These digital twins can be integrated in usual exploration workflows to support decision making. We do not advocate that non-invasive technologies will or should replace drilling but could largely contribute to a reduction of in-situ activities and therefore improve social acceptance.

Alteration Zonation Patterns Characterized by Hyperspectral Data within Orogenic Gold Deposits of the Ahafo South Camp, Sefwi Granite-Greenstone Belt, Ghana

Adu Agyapong, Jeffrey Bigelow, Michael Nemitz
Newmont Ghana Gold, Accra, Ghana

The application of hyperspectral analysis provides the ability to address inconsistent recognition of alteration minerals captured by traditional logging methods. Routine hyperspectral analysis of drill core provides semi-quantitative mineralogical data which allows for objective identification and classification of alteration assemblages and domains. The composition and distribution of alteration assemblages can be further refined by pairing the hyperspectral results with rock hardness measurements collected via a handheld rebound device. Within the Ahafo South camp, hanging wall alteration zonation developed within granodiorite and diorite host rocks, displays systematic mineralogical variation as a function of distance to known fluid pathways. Distal alteration is characterized by the conversion of amphibole to Mg-rich chlorite. With increasing proximity to the dominant fluid pathway, primary plagioclase and secondary chlorite are replaced by calcite and phengite. Continuing to approach the primary fluid conduit, Fe-carbonate and ammoniated mica form and phengite transitions to a muscovitic composition. Additionally, there is a shift in the chlorite 2250nm and 2350nm absorption features to longer wavelengths indicating more Fe-rich compositions before chlorite is ultimately altered to muscovite and dolomite. Proximal to fluid conduits within mineralized zones, the alteration assemblage consists of pervasive albite+dolomite+pyrite+gold±quartz with Fe-rich chlorite developed locally in mafic host rocks. This assemblage is characterized by complete destruction of primary textures and increased rock hardness. Though albite and quartz are not hyperspectrally active within the shortwave infrared range, their presence has been confirmed by XRD analysis and petrography and correlated to mineral proxy parameters generated from the hyperspectral data. Establishing a relationship between relative rock hardness and the semi-quantitative mineralogy generated from hyperspectral analysis provides a link to create consistent, multi-parameter, 3D domain models that can influence exploration targeting as well as mining and process related functions.

Collection of X-Ray Fluorescence Analyses of Reverse Circulation Drill Chips as a Tool in Geological and Alteration Modeling at the Castle Mountain Gold Deposit, California

Erik R. Tharalson[1, 2], Will Joseph[2], Oscar Vazquez[2], Wiley Skewes[2], Owen Nicholls[2], Thomas Monecke[1], Annelie Lundström[3], Egill Ö. Sigurpálsson[3]
1. Center for Advanced Subsurface Earth Resource Models, Colorado School of Mines, Golden, CO, USA, 2. Equinox Gold Corp, Henderson, NV, USA, 3. Minalyze AB, Sävedalen, Sweden

Castle Mountain is a Miocene-aged, disseminated-style, low-sulfidation epithermal gold deposit located in southern California. The deposit currently hosts reserves of 4.17 million ounces of gold at an average grade of 0.51 g/t gold. Mining operations resumed in 2020 after a 16-year hiatus. Phase 1 of mining is expected to produce approximately 25,000 ounces of gold per year for five years from backfill material, while phase 2 average annual production from bedrock will be approximately 200,000 ounces of gold over 14 years.

The host rocks include coherent lavas and volcaniclastic rocks that are intermediate and felsic in composition. Alteration associated with the mineralization has primarily resulted in increases of Si and K while areas outboard of the main mineralization show less significant compositional changes. The large deposit area, wide host rock variability, and inconsistency between multiple generations of logging geologists have proven to be significant challenges to building a meaningful 3D alteration model based solely on visual logging.

To refine the existing geology and alteration models, 8,205 meters of reverse circulation drill chips were analyzed using an automated X-ray fluorescence (XRF) scanner manufactured by Minalyze AB. The instrument enabled rapid (i.e., 60 s per chip compartment) collection of geochemical data that can be integrated into the geology and alteration models for the Castle Mountain deposit.

To date, results show that the coherent volcanic rocks in the deposit area are easily identified using immobile element ratios (e.g., Ti/Zr). Alteration zoning as defined by variations in the K/Al ratio show that more intensely altered rocks occur in proximity to the higher-grade areas of the deposit and contain the highest average gold grade. These initial results show that routine, rapid XRF analyses on reverse circulation dill chips can be pivotal to interpreting and modeling the large alteration zones around disseminated-style low-sulfidation epithermal deposits.

Getting the Most from Lab and pXRF Multi-element Geochemistry with Examples from the Red Lake Gold Complex, Ontario, Canada

Ned Howard[1], Vittoria Smith[2], Ryan Desjarlais[2], Allan Bieber[2]
1. Evolution Mining Ltd, Sydney, NSW, Australia, 2. Red Lake Gold Mine, Evolution Mining Ltd, Red Lake, ON, Canada

Emphasis is often placed on innovative technologies to improve both productivity and probability of discovery in mineral exploration. However, some of the highest impact improvements can be realised simply through better application and integration of existing techniques and technologies, often utilising datasets that have already been collected but not used.

A good example comes from a pre-existing, very large multi-element geochemical dataset from the Red Lake gold complex, Ontario, Canada, has been used for more accurate identification of lithological units and geological modelling in resource estimation and near-mine exploration targeting. Despite limitations due to significant analytical issues and a diversity of analytical techniques and elemental suites, the most stratigraphically important rock units can be discriminated using this dataset, including different geochemical suites of visually similar basalts and of felsic igneous rocks. This is allowing better geological modelling of complicated ore zones for improved resource estimation. The improved stratigraphic resolution also enables better reconstruction of mineralisation-related litho-structural architecture and better prediction of trap sites for improved near-mine targeting. Additionally, most stratigraphically important rock units at Red Lake can be more effectively discriminated with fit-for-purpose standards using pXRF analysis of drill core during logging, avoiding long turnaround times for traditional assay laboratory analysis.

While useful information can be gained from most historic geochemical datasets, this example shows that use of laboratories with sub-optimal analytical practices or misguided selection of analytical method, usually in pursuit of marginal cost savings, often undermines or destroys the potential value of historic and current geochemical analysis regimes.

SEG 2022 Conference: Minerals For Our Future

Application of High Resolution XRF Core Scanning in Early-Stage Mineral Exploration to Characterize the Distribution of Critical Metals

Hugh de Souza[1], Justin Daley[2], Mike Gunning[2], Alexander Seyfarth[3]
1. Geochemical consultant, Toronto, ON, Canada, 2. VR Resources Ltd., Vancouver, BC, Canada, 3. SGS, Janesville, WI, USA

While XRF continuous core scanning instruments have been available since the early 2000's their, application in early-stage mineral exploration has been limited, especially in comparison to the adoption of hyperspectral core scanners with the use of VNIR-SWIR cameras to map alteration mineralogy as a vector to ore. However, not all mineralised systems are alteration driven, for which geochemical gradients should be considered instead as a vector to ore. Nowhere is this truer than in exploration for critical minerals such as lithium, vanadium, nickel, niobium and the rare earth elements (REE) where the petrogenetic evolution leading to metal concentration is complex, and the current mineral deposit models for formation are not as well developed as they are for metals such as copper and gold. The aim of this presentation is to use the Cu-Au exploration project of VR Resources at the Hecla-Kilmer multiphase alkaline intrusive system with carbonatite in northern Ontario to illustrate the application of the Minalyze high resolution XRF core scanning unit, and the utility of its multiple outputs in early-stage exploration. The initial results from the Minalyze system illustrate the internal details of broad zones of REE and critical metal mineralization, including niobium and lithium, hosted within an extensive sulfide-bearing hydrothermal breccia system and calc-potassic alteration facies (fenitization).

Hyperspectral Mineral Mapping with UAVs: Challenges and Opportunities for Minerals Exploration

Sam Thiele, Sandra Lorenz, Rene Booysen, Yuleika Madriz, Moritz Kirsch, Richard Gloaguen
Helmholtz Institute Freiberg, Helmholtz Centre Dresden Rossendorf, Freiberg, Germany

Uncrewed aerial vehicles (UAVs) have rapidly become integrated into the mining lifecycle, with applications in exploration, production and post-mining management. Although mostly used for photogrammetric surveying, a variety of additional sensors are increasingly being deployed. Of particular relevance to mineral exploration, these include geophysical instruments (e.g., magnetometers, radiometers) and imaging spectrometers (e.g., multi- and hyperspectral cameras), that can be deployed to rapidly and accurately map structure, lithology and alteration. Hyperspectral sensors are especially sensitive to subtle mineralogical changes that can guide exploration and mining operations, albeit in well exposed areas (e.g., cliffs, open-pit workings, mountains or coastal outcrops). In this contribution, we present an overview of our current workflow for collecting and correcting UAV hyperspectral data for geological applications, and outline some of the important caveats and challenges when deriving geometrically and spectrally corrected data in topographically complex environments. We emphasise the importance of three dimensional topographic data, collected using photogrammetric techniques, and highlight the potential of combined digital outcrop and hyperspectral remote sensing workflows. An open-source implementation of this workflow (hylite) is introduced, and current challenges identified. Specifically, we highlight the need for rapid, robust and easy to use tools for processing data in the field, to facilitate QAQC and optimised survey planning and targeting. Finally, we present several case studies that apply hyperspectral UAV data to advance exploration for primary and secondary raw materials.

Hyperspectral Imaging for Lithium Exploration and Mining

Laura Tusa[1], Cecilia Contreras[1], Kai Bachmann[2], Richard Gloaguen[3]
1. TheiaX GmbH, Freiberg, Germany, 2. Erzlabor Advanced Solutions GmbH, Freiberg, Germany, 3. Helmholtz-Zentrum Dresden Rossendorf, Helmholtz Institute Freiberg for Resource Technology, Freiberg, Germany

Raw materials from geogenic resources are essential to enable the global shift to renewable energy sources. The global demand for raw materials is, therefore, being intensified. Electrification is an example of this and represents a key part of the energy transition through the decarbonization of the transportation sector. For this purpose, lithium is a critical metal due to its supply risk. However, significant hard rock lithium resources are found worldwide. The higher energy consumption for the extraction and concentration of lithium minerals from hard rock ores requires a thorough understanding of the mineralogical variability. This knowledge can be used for the selection of new technologies for pre-concentration.

Hyperspectral imaging has proven an excellent tool for providing quantitative data on mineralogy and microfabric attributes at a high spatial resolution in a fast, non-destructive and reproducible manner. Most lithium minerals have diagnostic signatures in either the short-wave infrared (SWIR) or long-wave infrared (LWIR) regions of the electromagnetic spectrum making their direct identification possible using hyperspectral imaging techniques. Zinnwaldite is an example of a lithium-bearing mineral present in high contents in the transitional Cínovec-Zinnwald deposit marked by the occurrence of numerous Li (Sn-W) bearing greisen bodies. Although the greisen bodies are mostly oriented sub-horizontally to the contact between the Cínovec-Zinnwald granite stock and its volcanic host rocks, they are irregularly intercalated with albite-granite, which is always Li- and generally Sn-poor. Large scale exploitation of the greisen will thus lead to variable amounts of dilution by albite granite. The fusion of hyperspectral drill-core imaging, automated mineralogy and geochemical data enabled the calibration of lithium grade at high resolution, allowing for the mapping of centimeter thick greisen layers which could be separated from the albite-granite using optical sorting with SWIR sensors.

Hyperspectral Core Imaging: Advanced Modern Exploration Techniques at the Toiyabe Exploration Project

David Browning, Steve Koehler, Savannah Reyes
Westward Gold, Vancouver, BC, Canada

The Toiyabe Gold Exploration project is located approximately 120 kilometers southwest of Elko, NV in Lander County, Nevada. The project hosts an exposed carbonate window consisting of the Devonian Horse Canyon and Wenban Formations. Exploration in the area dates to the 1960's with various drill programs focusing on shallow gold mineralization (primarily <180m). More recent drilling has identified a deeper zone of gold mineralization between 200-300m from surface associated with thrust faults and other compressional structures.

In 2021, approximately 3,050m of drill core and 11,900m of RC chips from previous drill campaigns were imaged using hyperspectral Visible-Near Infrared (VNIR), Shortwave Infrared (SWIR), and Longwave Infrared (LWIR) cameras. The data were processed to create individual mineral presence, mineral chemistry, and mineral texture images of the historic drill samples. The results of the imaging revealed important alteration patterns and textures related to large-scale alteration and structural features associated with the deeper zone of mineralization. Silicification, decalcification, and argillization are present with varying degrees of strength within drill intercepts of anomalous gold and the associated thrust fault, as well also occurring as haloes surrounding the mineralizing structure.

Hyperspectral core imaging has been used to quantify the degree of alteration across the previous drilling to identify series of alteration types associated with mineralizing structures. Using this "spectral fingerprint" as a vector for mineralization has led to the identification of additional areas of interest across the property that were not adequately drill tested. The application of this emerging analytical technique is aiding the refinement of existing drill targets from previous exploration efforts and the creation of new step-out targets that will be tested in the 2022 exploration campaign.

The Use of Hyperspectral Remote Sensing for Mineral Exploration in Italy: the Punta Corna Co-Ni Vein System (Piedmont, Italy)

Rita Chirico[1], Nicola Mondillo[1, 2], Carsten Laukamp[3], Simone Zanin[4]
1. University of Naples Federico II, Naples, Italy, 2. Natural History Museum, London, United Kingdom, 3. CSIRO Mineral Resources, Perth, WA, Australia, 4. Altamin Ltd, Oltre il Colle (BG), Italy

The aim of this study is to test the effectiveness of hyperspectral remote sensing for producing mineral distribution maps of the Punta Corna Cobalt Project, a brownfield exploration prospect owned by Altamin Ltd., located in the Servin Valley in the Western Alps (Italy). The objective of the work is to highlight the relative abundances and compositions of Visible Near (VNIR)- to Short Wave Infrared (SWIR)-optically active alteration minerals, genetically associated with hydrothermally driven metallogenetic processes. Hyperspectral remote and proximal sensing of mineral alteration footprints could support the mineral exploration in the area. The Punta Corna district is defined by hydrothermal polymetallic ores characterized by Fe^{2+}-rich carbonates and Co-Ni sulfide mineralization in form of E–W-trending subvertical veins, showing a maximum thickness of 6–7 m. Major veins formed during the post-metamorphic structural evolution of the Alpine belt, and sharply crosscut metaophiolitic host rocks' foliation. The country rock is affected by intense sericite–quartz-carbonate alteration proximal to ore. The method is based on mapping mineral occurrences through band ratios and feature extraction indexes performed on satellite images, using the PRISMA hyperspectral imagery (Italian Space Agency), combined with products acquired with other multispectral instruments, and supported by field-based hyperspectral data on hand specimens and DEM-based geomorphology. PRISMA provides imagery in a continuum of 234 spectral bands in the VNIR to SWIR wavelength regions, at a spatial resolution of 30 m. Laboratory IR spectroscopy and mineralogical (XRD) and geochemical (ICP-MS/ES) analyses were performed to identify possible target alteration minerals and to validate the interpretation of satellite images. The first results of the investigation, validated through fieldwork, show that satellite imagery is effective in mapping alteration minerals associated with the Co-Ni mineralization (i.e. white mica, chlorite, and supergene goethite), and could represent an additional tool for guiding toward prospective areas where Co-Ni-bearing veins occur.

Multi-source Hyperspectral Imaging of Drill-cores for the Exploration of Sedimentary Base-metal Deposits

Moritz Kirsch[1], Léa Géring[1], Samuel T. Thiele[1], Sandra Lorenz[1], Laura Tusa[2], Isabel C. Contreras Acosta[2], Andréa De Lima Ribeiro[1], Roberto De La Rosa[1], Houda Saffi[1], Richard Gloaguen[1]
1. Helmholtz-Zentrum Dresden-Rossendorf, Helmholtz-Institut Freiberg für Ressourcentechnologie, Freiberg, Germany, 2. TheiaX GmbH, Freiberg, Germany

The use of efficient and minimally invasive exploration technology in the mining sector is integral to sustaining the supply of raw materials in the future. Hyperspectral imaging is a spectroscopic technique that provides fast, high resolution, spatially continuous, and non-destructive mapping of mineralogy along drill cores. In this contribution we present a novel hyperspectral drill core scanning setup and processing workflow which we apply to mineralised cores from the Central European Kupferschiefer, one of the largest sediment-hosted accumulations of copper ores worldwide. For data acquisition, we use a Specim SisuROCK drill core scanner setup covering the visible-near infrared, shortwave, midwave and long-wave infrared spectral regions. The raw data are processed in near-real time using a custom-built processing pipeline that automatically identifies new datasets, applies sensor corrections and then co-registers the hyperspectral data from each sensor. Hyperspectral visualisations can then be virtually projected on the core using augmented reality to facilitate core logging activities and guide sampling. Innovative machine learning approaches are used for upscaling automated quantitative mineralogical data and lithological domaining. In our case study from the Spremberg-Graustein Cu-Ag deposit in Germany, we identified several spectral features associated with the occurrence of iron oxide, sulphates, kaolinite and carbonates. Whereas the iron oxides mark the oxidation along the path of the metal-bearing hydrothermal fluid in the sandstones immediately below the Kupferschiefer, Fe-rich carbonates and kaolinite are interpreted to represent products of hydrothermal alteration, as they delineate the position of high grade mineralisation. Changes in carbonate mineralogy also reveal fluid flow along styolites in carbonate rocks occurring stratigraphically above the main mineralized zone. The study suggests that hyperspectral imaging techniques are well suited to track hydrothermal fluid flow paths and vector towards base metal deposits in sedimentary basins and highlight the potential for real-time hyperspectral data acquisition and processing during drilling campaigns.

Hyperspectral Imaging for Mapping Outcropping Li-bearing Pegmatites

René Booysen[1, 3], Sandra Lorenz[1], Samuel Thiele[1], Yuleika Madriz[1], Warrick Fuchsloch[2], Paul Nex[3], Timothy Marais[2], Richard Gloaguen[1]
1. Helmholtz Institute Freiberg for Resource Technology - HZDR, Freiberg, Germany, 2. AfriTin Mining, Johannesburg, South Africa, 3. University of the Witwatersrand, Johannesburg, South Africa

The transition towards a net-zero economy has led to an increased need for several critical raw materials required for green technologies. Lithium (Li) is one of these materials, which is experiencing a sharp increase in demand that recycling alone is not capable of meeting. Significant mineral exploration using sustainable, efficient and innovative methods is required to not only improve mineral detection and mapping, but also foster social acceptability for the mining and exploration industry. Hyperspectral imaging (HSI) is a rapidly developing technology that allows for fast and systematic identification of key minerals and provides information about mineral abundances and associations. In this contribution, we propose an innovative approach for exploration by using hyperspectral data from multiple sensors at various scales to non-invasively map ore and pathfinder minerals. We showcase this approach at the Uis pegmatite complex located in Namibia. Hand-samples taken at the Uis pit were used to identify the relevant minerals as well as for training/validation purposes, and we acquired HSI data of the pit using a short-wave infrared (SWIR) camera mounted on a tripod. Using machine learning and computer vision techniques, we upscaled this information to the outcrop scale and created a three-dimensional (3D) point cloud of the pit with HSI attributes to map relevant Li-bearing minerals, such as cookeite and montebrasite. Results were validated using drill-core data, LIBS measurements and XRF analyses. We recently acquired uncrewed-aerial vehicle (UAV)-based SWIR data, to allow flexible data acquisition and mitigate access limitations. With in-house tools, the data is being processed (e.g., geometric and radiometric corrections) and we expect to map Li-bearing minerals in a similar manner. This approach enables rapid and efficient mapping of complex terrains in a sustainable exploration scheme, and can be used for monitoring and optimisation of ore extraction.

Mineralogical and Geochemical Vectoring Techniques in Advanced Argillic-Altered Rocks of British Columbia

Farhad Bouzari[1], Robert G. Lee[1], Craig J. Hart[1], Bram I. van Straaten[2]
1. MDRU-Mineral Deposit Research Unit, The University of British Columbia, Vancouver, BC, Canada, 2. British Columbia Geological Survey, Vancouver, BC, Canada

Zones of advanced argillic-alteration occur in the upper parts of porphyry systems, and where preserved may create the largest near-surface footprint of the hydrothermal system. Mineral exploration within advanced argillic zones is difficult because of their large size, the intense nature of the alteration, and the subtle mineralogical changes that can be challenging to identify.

This study characterizes alteration-mineral assemblages and compositions across zones of advanced argillic-alteration in three British Columbia properties: Tanzilla, Alunite Ridge, and Kemess North. A total of 230 rock samples were characterized by field observations, hyperspectral shortwave infrared, petrography, cathodoluminescence, scanning-electron microscope, and whole-rock lithogeochemistry.

Aluminum-rich minerals such as andalusite, corundum, diaspore, and topaz occur within zones of intense alteration and silicification with pyrophyllite and/or muscovite-clay assemblages. The muscovite color changes laterally, with respect to the main up-flow zone, to pale-green and green, in conjunction with a change to phengitic composition. A new K-Mg-Al molar diagram is introduced to map white muscovite-clay-(aluminum-rich minerals), pale-green muscovite, green muscovite-chlorite, and chlorite-dominated alteration assemblages. A K/Rb-Al diagram distinguishes the aluminum-rich phases from those of the muscovite-clay assemblages. Cathodoluminescence study indicates that quartz within more central parts of advanced argillic alteration displays red luminescence and typically occurs with blue luminescent pyrophyllite and clays.

Two trace element indices, MPIx (MDRU Porphyry Index) and MPIx-Lateral, are used to map the mineralization footprints vertically and laterally, respectfully. These indices provide tools to compare prospects in a district, identify size potential and level of exposure, and help target deep drilling. A Na-Ca depletion index maps the intensity of muscovite and advanced argillic-alteration, adding more new tools to define vectors for mineralization.

These research advances provide a toolset that can aid rapid and cost-effective exploration vectoring for porphyry copper and related epithermal mineralization within and around regions with advanced argillic-altered rocks.

Assessment of Tourmaline Composition as a Vectoring Tool for Sn-W Deposits – the Góis-Panasqueira-Segura Belt (Central Portugal)

Miguel Gaspar[1, 2], Isabel R. da Costa[1], António Mateus[1, 2], Ivo Martins[1, 2], Pedro Rodrigues[1, 2]
1. Departamento de Geologia, Faculdade de Ciências, Universidade de Lisboa, Lisboa, Portugal, 2. Instituto Dom Luiz, Faculdade de Ciências, Universidade de Lisboa, 1749-016 Lisboa, Portugal, Lisbon, Portugal

Tourmaline is a common accessory mineral in Variscan S-type granites and their metasedimentary host rocks and is often present in hydrothermal alteration haloes associated with Sn-W deposits, where tourmalinization can locally develop tourmalinites. About 2000 Electron Probe Micro Analyser (EPMA) data were compiled for tourmaline from the Góis-Panasqueira-Segura Sn-W belt, to assess its potential as a vectoring tool for Sn-W deposits. Granite tourmalines are typically schorlitic with low #Mg ($Mg/(Mg+Fe_T)$) and #Ca ($Ca/(Ca+Na+K)$), whereas metasediment tourmalines are dravitic with higher #Mg and #Ca. #Mg *vs.* #Ca plots highlighted: magmatic differentiation trends for composite granites, with progressive decrease in #Mg and #Ca; compositional variability within metasediments; and mixing lines, for each individual hydrothermal system, between tourmaline compositions buffered by metasediments and by magmatic-hydrothermal fluid sources. The mixing line spread discloses the hydrothermal system magnitude (time, space, and mineralizing fluid volume), as shown by the wider trend for the Panasqueira deposit when compared to smaller deposits such as Góis or Segura. However, full interpretation of mixing lines, or deviations from it, require detailed petrography and careful microprobe spatial and compositional control, to address aspects such as multiple tourmaline generations, compositional zoning, or even diffusion mechanisms. In general, hydrothermal tourmaline documents a progressive increase in #Mg and #Ca reflecting a concomitant rise of the magmatic-hydrothermal fluid component. Except for Panasqueira, tourmalinites at the Penamacor-Monsanto batholith exocontact record the lowest #Mg values denoting its proximity to the intrusion. For systems like Góis, where no granites are known in the vicinity, the trend starts at higher #Mg values and a more distal position relative to a causative granitic intrusion is believed. Tourmaline composition by EPMA, a cost-effective routine method, holds promise as an exploration tool for Sn-W deposits.

This is a contribution by FCT/UIDB/50019/2020 grant and of ERA-MIN/0002/2019 project (MOSTMEG)

Assessment of Wolframite-Scheelite Replacement as a Vectoring Tool In Sn-W Lode Deposits – a Case Study of the Argozelo Mine (NE Portugal)

Pedro Francisco[1], Miguel Gaspar[2], Pedro Santos[3]
1. Colorado School of Mines, Golden, CO, USA, 2. Faculdade de Ciências da Universidade de Lisboa/Instituto Dom Luiz-FCUL, Lisbon, Portugal, 3. Empresa de Desenvolvimento Mineiro, SA, Lisbon, Portugal

The tin-tungsten ore deposit of Argozelo is a lode mineralization located in the NE sector of the Central Iberian Zone (CIZ) and included in the "Iberian Sn-W metallogenetic province". The Sn-W quartz veins are hosted in upper Devonian CIZ metasedimentary rocks. Due to the existence of several productive granitic intrusions at a regional scale, the lode mineralization is believed to be related with a deep, hidden granitic body. Systematic core sampling, petrographic studies, and extensive microprobe analysis helped define the main ore vein phase, mineral assemblages, paragenetic sequence and a preliminary hypothesis for the chemical evolution of the magmatic-hydrothermal system. Common features of Argozelo's ore veins include wolframite-scheelite replacement textures and the presence of euhedral fluorapatite selvages. The electron microprobe study of wolframite, scheelite and co-precipitated fluorapatite within the ore veins showed that: wolframite precipitates dominantly as hübnerite ($Mn^{2+}WO_4$); scheelite replaces wolframite by partial to total dissolution followed by reprecipitation; an early fluorapatite generation rich in Mn and Fe, and poor in Ca (Ap1) occurs with the wolframite stage, and a late Ca-rich and poor in Mn and Fe fluorapatite (Ap2) precipitates with the scheelite stage. Thus, mineral chemistry documents a progressive increase in the Ca/(Fe+Mn) ratio during the magmatic-hydrothermal fluid evolution. Since regional calc-silicate units are considered the primary Ca source, the increase in the Ca/(Fe+Mn) ratio should reflect the input of an external fluid component resulting from a progressive higher fluid-wallrock interaction. Intense wolframite-scheelite replacements, with predominance of hübnerite and Ca-rich fluorapatite (Ap2), also reflect distal deposit features, whereas fewer and less extensive wolframite-scheelite replacements, with prevailing ferberite ($Fe^{2+}WO_4$) over hübnerite and fluorapatite (Ap1) higher in Mn and Fe, suggest proximity to a causative granitic intrusion.

Leaving a Positive Legacy, Making a Difference in Africa

Robert Kaemba
First Quantum Minerals, Ndola, Zambia

The reliance towards natural resources has an even greater importance now more than ever. This reliance has far reaching applications due to the varied use and applications in green energy, technology, infrastructure and electrification. As we transition towards a more carbon-neutral economy, there is an increased emphasis on production and expansion of existing projects as well as the discovery of new projects, to bridge the potential supply gap of critical metals.

An increase in demand of this nature must be aligned to social, environmental and political expectations. Considerations must be made to the ways in which responsible Mining can be achieved in a Sustainable manner. This includes prospecting, exploration, construction, expansion, abandonment, decommissioning and repurposing. This can yield a number of direct and indirect social, economic and environmental impact.

As a result, Social License to Operate is becoming an essential tool. This can be achieved by greater community engagement as a sustainability strategy, and deserves attention. My talk explores the social, economic and environmental impact created by industry in Africa, with a focus on aspects that may be obscure from our usual line of sight as Scientists.

If an approach using a key set of "Golden rules" is applied, this can foster and lead to the potential positive societal and environmental impact that can achieve a long lasting positive legacy.

Metal Sourcing for a Sustainable Future

Richard Gloaguen[1], Ortwin Renn[2], Cathryn MacCallum[3], Jon Russill[4], Moritz Kirsch[1], Leila Ajjabou[1]
1. Helmholtz Institute Freiberg for Resource Technology, Freiberg, Germany, 2. Institute for Advanced Sustainability Studies, Potsdam, Germany, 3. Sazani Associates, Carmarthen, United Kingdom, 4. SRK Exploration Services, Cardiff, United Kingdom

Drastic measures are required to meet the standards of the Paris Agreement and limit the increase of global average temperatures well below 2°C compared to pre-industrial levels. Mining activities are typically considered as unsustainable but, at the same time, metals such as cobalt and lithium are essential to sustain the energy transition. Indeed, several sustainability goals defined by the United Nations (UN) require large quantities of raw materials. Exploration and extractives activities are required in order to contribute to meeting sustainability standards. Future sourcing of metals will need to implement procedures that go well beyond current ecological, economic, and social requirements and practices. In order to comprehend the barriers to acceptance we assess the usual sustainability criteria and how they apply to the extractives sector. Sustainability can only be achieved if one accepts that the natural capital can be substituted by other forms of capital (so called weak concept of sustainability). Sourcing the raw materials increasingly demanded by our societies will need transparent and inclusive stakeholder participation as well as a holistic understanding of the impact of mining activities to reach this weak sustainability status. Regardless which governance measures are introduced there will always be trade-offs between opportunities such as facilitating the energy transition or e-mobility on the one side and risks for ecosystem and social communities on the other side. Our analysis shows that the sustainability of mining cannot be reached without harmonized political instruments and investment policies that take the three pillars of environmental, economic, and social sustainability as a major priority. This submission results from research conducted within the EU funded project INFACT.

A New Paradigm for Responsible Exploration and Sustainable Mining

Gregory Wessel[1], Etzigueri Góngora Ubeda[2], Ernesto O. Cordero[3]
1. Geology in the Public Interest, Vashon, WA, USA, 2. Universidad Nacional Autónoma de México, Instituto de Geología, Mexico City, COA, Mexico, 3. Saint Paul University / University of Ottawa, Ottawa, ON, Canada

The principles of responsible mining have evolved over the last few decades, following growing realization that mining as it was done in the past led to permanent degradation of the surrounding environment, health impacts to nearby residents, and regional impacts to air and water quality that were difficult if not impossible to mitigate. People both inside and outside the mining industry found themselves having to reckon with costs of mining and challenges that had always been there but were never counted.

Until recently, exploration in advance of mining was driven by the same low standards to which mining had been held in the past. The challenge for explorationists today has shifted from finding high-grade deposits or large bulk-minable deposits to deposits that can be exploited while still meeting all the requirements for responsible mining. This new challenge requires skills and knowledge that are only starting to be included in geoscience education, including the importance of social license and ethical and moral constraints that address the environment, nearby residents, and future generations.

Standards now exist against which the actions of explorationists can be measured to ensure that subsequent mining is responsible and of benefit to all parties, including future generations. Organizations like the Initiative for Responsible Mining Assurance and the International Council on Mining and Metals offer guidelines for responsible mining that translate directly to guidelines for responsible exploration. Exploration programs that are endorsed by these organizations exhibit a commitment to honor social and cultural contexts, eliminate environmental impacts and remediate past damage. By responsibly self-regulating, reducing carbon emissions, and planning for full restoration, mining can even approach a degree of sustainability and become a real force for positive change. To do this, responsible and ethical exploration and mining standards must be implemented from the very start.

Predicting the Impact of Mining on Watersheds: a Case Study from the Ocoña Watershed in Arequipa, Peru

Isaac Simon[1], Katharina Pfaff[1], Alexis Navarre-Sitchler[1], Elizabeth Holley[1], Gary Vanzin[1], Jorge Crespo[1], Madeleine N. Guillen Gomez[2], Sergio Ticona Corrales[2], Jonathan Sharp[1]
1. Colorado School of Mines, Golden, CO, USA, 2. Universidad Nacional de San Agustin de Arequipa, Arequipa, Peru

With increasing resource development comes associated responsibility to prevent environmental contamination. A variety of ore deposits are mined in the state of Arequipa in Peru, where artisanal and small-scale mining serve as a major source of livelihood to the local community. Additionally, both local and international companies are conducting brownfields and greenfields exploration throughout this region. However, mining in Arequipa also serves as a point of contention among citizens due to concerns of contamination of local freshwater sources. These concerns have been particularly documented in the Ocoña watershed, where mining co-exists with agriculture and fishing along the river valley.

The aim of this study is to better understand the impact of mining activity, the buffering capacity of the local geology and water rock interaction on water quality. Principle component analysis of Ocoña river water data suggests a contribution of Cu, Mn, Al and Fe in areas where mining is the dominant anthropogenic activity. Water discharge-concentration relationships are indicative of greater contribution of such metals at higher river discharge levels. Despite flushing of Cu-Mn-Al-Fe at higher discharge, the neutral to basic water pH conditions (median pH = 8.09) maintains concentrations of such metals at lower levels, compared to the neighboring Tambo and Chili watersheds. Local geologic characteristics in mining areas that are capable of neutralizing acidic solutions include vein carbonate and propylitic-altered rocks.

The environmental impacts of resource development are a critical consideration in early stages of mineral exploration and ultimately mining. Understanding the relationships of mining, geology and the Cu-Mn-Al-Fe hydrogeochemical signature can be used to predict risks of contamination of local watersheds prior to any exploration activities. Ultimately, knowledge about the interplay of human activity, the local geology and its natural acidity buffering capacity can help minimize adverse water quality impacts and inform remediation efforts.

Coexistence Between Large-scale Mining (LSM) and Artisanal and Small-scale Mining (ASM) in Peru and Colombia

Oscar F. Rodriguez, Elizabeth Holley
COLORADO SCHOOL OF MINES, GOLDEN, CO, USA

Artisanal and small-scale mining (ASM) is common in mineral-rich locations in the developing world, generating a significant share of production for commodities such as gold. ASM has expanded due to economic and environmental pressures. Large-scale mining (LSM) companies will also need to expand to meet future demand for minerals and metal resources. The likelihood that these expansions will result in further points of contact between the LSM and ASM sectors is high. In this scenario, coexistence between ASM and LSM is key and should be seen as the capacity to build synergies and operate together. We built a database of sites in Colombia and Peru where ASM and LSM both occur, documenting the technical characteristics such as deposit and mineralization type, mining methods, recovery process and the social context including ethnicity, demography, economic activities, and social organiztions. We observed that coexistence occurs between ASM and both junior and senior LSM companies, covering all the stages of the mining cycle. Coexistence scenarios are most common in gold projects but cover a wide variety of mineralization and deposit types including epithermal, porphyry and VMS among others. We conclude that the use of ASM activities by LSM companies as exploration vectors is the main cause of convergence between LSM and ASM in developing countries. Governments have taken significant efforts to create tools for LSM companies and holders of mining titles to reach agreements with ASM miners. Nevertheless, the success of these agreements largely depends on the willingness of the parties. To reduce the risks of social harm in developing countries while reducing the social risk to the exploration and mining projects economic geologists must be aware of the different strategies that governments and LSM companies are implementing to address the coexistence in order to incorporate this knowledge in project evaluation and risk assesment.

Propagation of Bias: the Data Behind the Damage of "Do Your Time"

Stephanie E. Mills
Utah Geological Survey, Salt Lake City, UT, USA

Research shows that men often benefit from the perception of promise whereas women are judged on demonstrated experience (e.g., Player et al., 2019). In economic geology, this bias can be seen in the "do your time" mentality, where previous experience disproportionately benefits men for raises and career progression versus women and other minority populations (such as racial minorities and the LGBTQ+ community) who must demonstrate ability before receiving advancement opportunities. This inequity is particularly pronounced when it comes to fieldwork, a key component in many disciplines of economic geology. The "do your time" mentality is particularly damaging during the early-career stage, where it is also pervasive. Managers influenced by the "promise vs. demonstrated" bias are more likely to start women and minorities at lower wages and positions, while believing that the inequity will balance over a career—once they have "proven" themselves. Presented here are data from professional surveys and individual examples demonstrating the opposite, that these early career inequities propagate throughout careers resulting in increasing salary deficiencies and consistently delayed career progression, meaning women and minorities must work years longer than their counterparts to achieve the same income or career level, or may never be able to do so. The propagation of "do your time" bias from early career can be so dramatic as to destabilize intergenerational elasticity and negatively impact the socioeconomic status of the next generation, particularly in modern families that are dependent on two salaries or where a woman or minority is the sole earner. This propagation of bias undermines the narrative that exploration and mining create high paying jobs. Solutions are discussed to improve oversight on hiring decisions, even for the most junior positions.

Player et al., 2019, Overlooked leadership potential, Frontiers in Psychology, v. 10.

Exploring New Business Models for Mineral Exploration

Martha Henderson[1], Luís Chambel[2]
1. Uppsala University, Visby, Sweden, 2. Sinese, Lisbon, Portugal

Currently, it's not worthwhile to invest in exploration:

- With only 1 project becoming a producing mine out of 500 to 1000 grassroots exploration projects (Allen, 2015; Eggert, 2010), more money is invested than value generated by discoveries (Schodde, 2020).
- The industry faces increasing difficulty obtaining the social license to operate as well as rigorous environmental standards and permitting requirements.
- Mineral deposits are deeper and located in more remote areas (with higher costs and lower success rate). Also, the ores are more complex to process and have lower grades (requiring higher specific energy, water and land uses in mining).

These challenges occur at a time when new mines are needed to transition to a sustainable economy. Green energy, high-tech and defense industries demand a growing volume and variety of raw materials, such as lithium, cobalt, magnesium, copper and REE. The aftermath of the coronavirus and the geopolitical long-term consequences of war in Ukraine exacerbate the trend to deglobalization, each block aiming for self-sufficiency in critical minerals.

This creates both opportunity and responsibility for the exploration industry; yet it must reinvent itself. It must be more efficient, with better targeting, evaluation tools, and methodologies. It needs to fundamentally change its business models, focusing on natural and social systems rather than only on mineral deposits, shareholders, and Boards, while transforming from a linear to a circular economy business model.

We look at the operation, business strategy and value creation of a sample of junior mining companies listed on the TSX-V. A vast majority of junior companies is focused on gold exploration, adopting a business-as-usual approach. However, recent years have witnessed an increased number of new ventures focusing on lithium, and other battery and high-tech metals, some of them with innovative business models.

High Temperature (>800°C) Brine and Sulfide Melt Interaction During the Formation of Northern Bushveld Magmatic Sulfide Cu-Ni-PGE Deposits

Katie McFall[1], Iain McDonald[2], Marina Yudovskaya[3], Judith Kinnaird[3], Jacob J. Hanley[4], Mitchell Kerr[4], Brian Tattitch[5]

1. UCL, London, United Kingdom, 2. Cardiff University, Cardiff, United Kingdom, 3. University of the Witwatersrand, Johannesburg, South Africa, 4. Saint Mary's University, Halifax, NS, Canada, 5. Centre for Exploration Targeting, University of Western Australia, Perth, WA, Australia

The role of volatiles in the development of layered intrusion-hosted magmatic sulfide deposits is still poorly understood. We present petrological and fluid inclusion evidence that an aqueous brine phase was present during the formation of Ni-Cu-PGE mineralisation throughout the northern Bushveld Complex, which hosts some of the largest Ni-Cu-PGE orebodies on Earth. This brine phase was present while the system was at least partially molten and we present direct evidence of brine – sulfide melt interaction.

Fluid inclusions were identified in cumulate magmatic silicates (feldspar, olivine, and pyroxene) in the Aurora, Waterberg PTM, Troctolite Unit and Platreef deposits. These comprise brine and vapour inclusions which do not crosscut crystal boundaries. Optical microscopy and confocal Raman spectroscopy show that the brine inclusions contain multiple daughter minerals (including halite and carbonates), 10-15% vapour and 15-25% liquid water; whereas vapour inclusions contain CH_4 and N_2. Microthermometry shows that brine inclusions homogenise to liquid by vapour disappearance between 819 – 1000°C in the Aurora deposit (n=281), between 877 – 994°C in Waterberg (n=256), between 860 – 942°C in the Troctolite Unit (n=428) and between 829 - 988°C in the Platreef (n=227). Halite dissolves between 521 – 697°C, giving salinities of 61-87 wt.% NaCl equivalent.

Examples of co-eval interaction and entrapment of brine and Cu-rich sulfide melt are observed in all deposits, implying this was a common factor during their development. The PGE in the Aurora project and the T zone at Waterberg are hosted in platinum group minerals (PGM) that are spatially removed from magmatic sulfides and hosted in late silicates such as quartz, suggesting PGE remobilisation by hydrothermal processes. The petrological similarity between inclusions in all Northern Limb deposits suggests the high temperature fluids had a similar origin, either from late magmatic degassing or possibly from volatiles released from assimilated dolomite country rock.

Epithermal Gold Mineralization Hosted in an Ophiolite: a Case Study of the Malabeg Prospect, Zambales Ophiolite Complex, Philippines

John Emmanuel S. Fungo[1], Jillian Aira S. Gabo-Ratio[1], Rosalito A. Alonso[2], Kyoichi Ito[3], Pearlyn C. Manalo-Balmater[4], Karl D. Jabagat[5], Marco Alfredo J. Barrientos[1], Kotaru Yonezu[6]
1. University of the Philippines-National Institute of Geological Sciences, Quezon City, Philippines, 2. Cordillera Exploration Co. Inc., Taguig City, Philippines, 3. Sumitomo Metal Mining Company, Tokyo, Japan, 4. Akita University, Akita, Japan, 5. National Chung Cheng University, Chiayi, Japan, 6. Kyushu University, Fukuoka, Japan

An intermediate sulfidation epithermal mineralization was recently recognized in the Malabeg Prospect situated within the Zambales Ophiolite Complex and flanked by the Bataan Volcanic Arc in Luzon Island. NW-trending quartz-adularia veins are hosted in the serpentinized harzburgite and diabase dikes of the Cabangan Massif. Gold-bearing veins and floats were discovered to have grades of up to 74 ppm gold.

The alteration and mineralization stages were determined through petrography, XRD and SEM-EDX. Stage 1 consists of talc, pyrophyllite and magnesite, which were produced from alteration of the serpentinized peridotites. Stage 2 consists of carbonates such as magnesite and dolomite at the rims of and as veins cutting Stage 1 talc-carbonates. Stage 3 consists of quartz-carbonate veins. Arsenopyrite is present in both Stages 2 and 3, wherein it cuts both quartz and calcite. Stage 4 consists of quartz-adularia veins where massive, comb and brecciated textures were observed. Ore mineralogy consists of chalcopyrite ± pyrite + galena + sphalerite ± electrum. Supergene alteration is represented by iron oxides, pyrolusite and secondary covellite.

Mean iron-sulfide mole percent for nine sphalerite points coexisting with pyrite and pyrrhotite is 14.91. Two-phase fluid inclusions were found in Stage 4 quartz-adularia veins. Modal homogenization temperatures range from 180-200 °C while salinity values range from 3.87-7.6 wt.% NaCl. Sulfide mineralogy, homogenization temperature, and iron-sulfide mole percent of sphalerite indicate an intermediate sulfidation state for the quartz-adularia veins of the Malabeg prospect.

The results suggest listwanitization of the serpentinized peridotites (Stages 1-3) through reaction with intermediate sulfidation fluids from a magmatic source possibly related to the Bataan Volcanic Arc. The high buffering capacity of the peridotites lowered the sulfidation state of the initial system, resulting to precipitation of arsenopyrite. The shallow depth of the listwanitized zone made it susceptible to brittle deformation, where subsequent IS fluids precipitated the quartz-adularia veins.

Fault-surface Map Restorations of San Manuel-Kalamazoo: Reassessing the Tilting History

Juan F. Fajardo[1], Eric Seedorff[2]
1. BHP, Tucson, AZ, USA, 2. University of Arizona, Tucson, AZ, USA

This study uses fault surface maps to determine the direction and amount of slipping on normal faults in the vicinity of the San Manuel-Kalamazoo porphyry Cu-(Mo) deposit, Arizona. Data from drill holes, crosscutting relationships, and stereonet rotations of bedding attitudes in synextensional sedimentary rocks permit estimation of the initial orientations of faults and the amount of tilting of the ore shell.

Normal faults can be assigned to six fault sets. Sets are based on orientation and relative age relationships: (1) Cloudburst, (2) Turtle, (3) San Manuel, (4) Black Canyon, (5) Cowhead Well, and (6) Cholla. The present-day dips of faults range from high-angle (Cholla, Black Canyon), to moderate-angle (San Manuel), to subhorizontal and overturned (Cloudburst).

Three additional dismembered segments of the Cloudburst fault were identified concealed beneath younger rocks, as well as three segments of the Turtle fault. The erosion surface that bounds the eastern side of the Purcell window and overlies the Kalamazoo deposit consists of three segments separated by inflections in dip. Successively deeper segments of the erosion surface dip more steeply to the east, at ~20°, ~30°, and ~60°. Both the presence of a 60°E-dipping segment of the erosion surface on the Purcell window and a small, gently-dipping segment of the Cloudburst fault in the footwall of the Turtle fault are compelling geologic evidence that the Turtle fault cuts and offsets the Cloudburst fault and does not constitute a lateral ramp to it. All faults contribute to tilting of the San Manuel-Kalamazoo ore shell, which has been tilted ~62°E, similar to dips of the oldest Tertiary rocks. The axis of the ore shell has a current trend of S57°W and a plunge of 20°SW; the original, restored axis trends S56°W and plunges ~82°SW, implying that the axis of the orebody was emplaced within ~8° of vertical.

SEG 2022 Conference: Minerals For Our Future

The Santo Tomás Deposit, Sinaloa, México: a Particularly Structurally Controlled Porphyry Cu-Mo System and Its Metallogical Implications

Jocelyn Pelletier[1, 2], Paul McGuigan[1, 3], William X. Chavez[4]
1. Oroco Resources Corp., Vancouver, BC, Canada, 2. Geogenius Exploration Consulting, Montreal, BC, Canada, 3. Cambria Metallogenics, Vancouver, BC, Canada, 4. New Mexico Tech University, Albuquerque, NM, USA

Abstract. Oroco Resource Corporation recently began exploring the porphyry Cu-Mo deposit at the Santo Tomás project, Sinaloa, México. A recent re-estimation of the historical mineral resource indicates 822 million tons with an average grade of 0.32% Cu. The 57.2 Ma Santo Tomás hydrothermal system is developed within Cretaceous-age carbonate rocks that overlie Jurassic-age andesite. Our work indicates that the deposit corresponds to a porphyry dyke complex; porphyritic monzonite intrusions can be followed at the surface for more than 4.5 km. In addition to porphyry-style mineralization, the Brasiles zone and the upper part of the Santo Tomás deposit hosts Zn-Pb-Fe-Ag-Au polymetallic skarn and Au-Ag specularite-mushketovite-quartz veins.

Santo Tomás differs from other porphyry copper-molybdenum deposits by its extended intrusion-related length and dyke-like morphology, but mainly for the Cu-stockwork localized in this inclined porphyry intrusion complex. Laminations are present in all the generations of veins that reach the upper part of the system, suggesting the interaction with meteoritic waters. Field geological observations indicate that a right-lateral component was active along local wrench faults that controlled the emplacement of the deposit. Due to this strong structural control, hydrothermal events are superposed along the same structures, reflected in the greatest copper grades occurring along specific vein directions; mainly the A-veins forming breccia when overprinting the biotite-rich potassic alteration zones. Andesite and limestone host rocks have been altered and mineralized because they acted as lithologic-geochemical buffers for mineralization, creating stockwork zones with locally greater copper grades. We surmise that this is attributable to structural porosity, resulting in wider veins, and reduced density of veins. To explain the abundance of multiples generation of laminated veins, we proposed a model-based of dynamic strike-slip and wrench fault setting that would have created more structural porosity, allowing magmatic fluids to reach a higher level in the system.

He, Ar, S, and Pb Isotopic Constraints on the Origin of the Shuanghe Gold Deposit, China

FengChun Li, Qingdong Zeng, Shuai Gao, RiXiang Zhu, Shaoxiong Chu
Institute of Geology and Geophysics, Chinese Academy of Sciences, Beijing, China

The Shuanghe gold deposit lies on the southern margin of North China Craton. Based on vein crosscutting relationships, mineral assemblages, and alteration characteristics, the formation of the deposit can be divided into four stages: 1) the barren quartz ± pyrite stage; 2) the quartz ± pyrite stage; 3) the quartz-polymetallic sulfides-ankerite stage; and 4) the quartz-carbonate stage. The abundant gold being mainly hosted at the stage 2−3. Pyrite-hosted fluid inclusions yield $^{3}He/^{4}He$ ratios of 0.05–0.25 Ra (most are greater than 0.14Ra; Ra = 1.4×10^{-6}) and $^{40}Ar/^{36}Ar$ ratios of 758.4–3560.1(average = 2024.7), which indicate the ore-forming fluids were predominantly crust-derived with minor mantle-derived components. The $\delta^{34}S$ values of pyrite range from – 4.2‰ to 2.1‰, show a relatively narrow range and consist with deep seated magma source. Lead isotope analyses indicate that the sulfides have $^{206}Pb/^{204}Pb$, $^{207}Pb/^{204}Pb$, and $^{208}Pb/^{204}Pb$ ratios from 18.226 to 20.016, from 15.597 to 15.880, and from38.475 to 39.396, respectively, indicating a mixed Pb sources with obvious proportions of mantle. Combined with previous data, we suggest that the ore-forming fluids and metals of the Shuanghe gold deposit stemmed directly from the regional mantle-derived magmatic hydrothermal system. The gold mineralization may genetically associate with the Early Cretaceous magmatism in an extensional tectonic setting, which is coupled with the large-scale lithospheric extension and thinning on the southern margin of the North China Craton.

Are Mafic Rocks a Source of Cu, Co, Ni and V in the Central African Copperbelt?

Zoe Phelps-Barber[1], David A. Holwell[1], Daryl E. Blanks[1], Robert Kaemba[2], Munshya Zimba[2]
1. University of Leicester, Leicester, United Kingdom, 2. First Quantum Minerals, Ndola, Zambia

Sediment-hosted stratiform deposits are significant sources of base metals globally, and with heightened demand in response to the development of low carbon energy production, the need to discover and exploit new deposits for these base metals is imperative.

Red-bed sequences in sedimentary basins are the predominant source of base metals within sediment-hosted deposits, however within the Central African Copperbelt (CAC), red-beds are not sufficiently voluminous to account for the metal endowment. The CAC hosts the majority of the world's Co resources, and a large share of the Cu resource alongside hosting some unusual hydrothermal Ni and V deposits, which leads to the question; what else could be a source of Cu, Co, Ni and V in the CAC? The basement and surrounding strata have been suggested as additional sources, however mafic bodies found throughout the CAC, which are inherently enriched in Cu, Co, V and Ni, have been overlooked and relatively uninvestigated.

Mafic bodies studied have undergone a minimum four stages of alteration, represented by nine different alteration styles, which have been classified through quantitative mineralogy. The textures and mineralogy suggest two different fluids enriched in Na, Ca, K and Cl, have mobilised Cu, Co, V, and the normally immobile Ni, to varying extents.

Through classifying alteration styles, geochemical proxies can be determined which can be applied to drillcore databases to determine the abundance of each alteration style. Through understanding the abundance of alteration, we are able to constrain the amount of Cu, Co, Ni and V leached from altered mafic rocks at a deposit scale. This has broader basin-wide implications to the significance of altered mafic rocks as a source of Cu, Co, Ni and V within the Central African Copperbelt.

Exploration for Sediment-Hosted Copper in Europe Using Mineral-system Analysis with Hydrocarbon Data

Howard C. Golden[3], William A. Heins[2], Peter Birch[1], Richard Jameson[1]
1. Getech, London, United Kingdom, 2. Getech, Kingston, NY, USA, 3. Getech, Perth, WA, Australia

Many critical mineral systems originate under specific genetic conditions, similar to hydrocarbon systems. We tested this concept by using hydrocarbon-focused geological information from proprietary databases to search for sedimentary copper deposits. The data are global and contain an extensive catalogue of data layers for every geological stage from the Pleistocene to the Asselian, at the beginning of the Permian.

This exercise used hydrocarbon exploration information rarely available to mining explorationists to attempt to predict where sedimentary copper deposits would form in the Kupferschiefer region of Europe. The database was interrogated to extract proxies for genetic criteria for Wuchiapingian (Permian stage that includes the Rotliegend and Zechstein formations) sediment hosted copper formation in Europe. The genetic criteria included depositional environment, copper source, copper-transporting brines, metal trap and fluid circulation.

The proxies for these genetic criteria included oceanic highstand-lowstand range; 170 km lithospheric thickness; oxidised continental sediments and/or volcanics; preserved evaporites or high Permian evaporation; total organic carbon; sedimentation rate; and basement heat production.

In this Kupfercshiefer Copperbelt study all the genetic proxy layers were incorporated into a GIS environment and analysed to determine where all or most of the criteria were co-located, resulting in a 'stoplight' system wherein green represents where all or most of the layers coincide, whereas yellow and red colours representing where fewer or none of the layers respectively are coincident.

The final stage of the experiment was to overlay known sedimentary hosted copper deposit locations on the stoplight map. In the Kupferschiefer case, the coincidence is excellent. The success of the test case bodes well for applying the same analytical technique to other geographical locations, other depositional ages, and other deposit types. Improving this analytical strategy and applying it is likely to highlight unexpected new search spaces for highly desirable commodities.

The Shale-hosted Las Cruces VMS, Iberian Pyrite Belt, Spain: Not a Brine Pool Deposit

Simon Jones[1], Tim Ireland[1], Matthew Leybourne[2], Eva Stüeken[3], Gil Machado[4]
1. First Quantum Minerals, West Perth, WA, Australia, 2. Queen's University, Kingston, ON, Canada, 3. University of St Andrews, St Andrews, United Kingdom, 4. Chronosurveys, Almada, Portugal

Las Cruces is a shale-hosted volcanogenic massive sulfide (VMS) deposit located in the southeast Iberian Pyrite Belt, Spain. The pre-mining resource (M+I) included 38.9 Mt of hypogene ore at 1.3% Cu, 1.2% Pb, 2.5% Zn, and 30.6 g/t Ag; and 14.2 Mt of supergene ore at 5.7% Cu. Previous studies concluded shale-hosted VMS deposits in the Iberian Pyrite Belt, including Las Cruces, formed via exhalation of saline, sulfur-deficient, metal-bearing hydrothermal fluids into the sea. These fluids were considered immiscible in seawater, and to have ponded in bathymetric lows forming reduced and acidic brine pools from which metal sulfide minerals precipitated.

Footwall rocks at Las Cruces are andesites and basalts that comprised a submarine caldera, overlain by a shale unit that thins over the volcanic edifice. A dense sulfide vein stockwork is present in the volcanic rocks and the shales. The shales also contain stacked, bedding-parallel, massive sulfide lenses concentrated in a middle, organic-rich sub-unit. Distal shales contain pyrite nodules with strongly negative $\delta^{34}S$ values indicative of bacterial sulfate reduction of seawater. Sulfide minerals within the deposit display positive $\delta^{34}S$ values, with massive sulfides averaging +6.2‰. Lithostratigraphy and palynology suggest the middle, organic-rich shale represents the ~358.9 Ma Hangenberg Mass Extinction Event. However, paleo-redox indicators including Mo, Fe/Al, and TOC all suggest conditions were weakly anoxic to oxic. Visual kerogen typing, C/N ratios, and $\delta^{13}C_{org}$ of the shale confirms mixed types of organic matter, but nitrogen stable isotopes ($\delta^{15}N$) show no evidence of abundant bio-available nitrogen venting as seen in some modern vents. None of the attributes of the shale are consistent with pockets of restricted water column circulation in a brine pool. Therefore, mineralization was most likely epigenetic with fluids derived from magmatic sulfur (~0‰) and seawater (~+20‰) in a ratio of ~2:1.

An Experimentalist View into Natural Carbonatite Complexes

Michael Anenburg
Australian National University, Canberra, ACT, Australia

Many silica-poor carbonatite complexes occur in a predictable geometrical form of concentric rings, with calcite carbonatite at the rim, followed by dolomite or ankerite carbonatites at the core. In this association, REE mineralisation typically occurs within the Fe-rich parts of the carbonatite, whereas the calcite carbonatite is mostly barren of REE. These observations mirror experimental studies in which a carbonatite melt will fractionate calcite followed by dolomite, leading to an increase in Fe and other incompatible elements (REE included), until LREE minerals such as monazite and burbankite precipitate in the Fe-rich residual melt. Burbankite is rarely preserved and is instead altered to aggregates of LREE minerals (commonly bastnäsite and other REE-fluorcarbonates) and baryte. Exceptionally Fe-rich oxidised rocks that occur in carbonatite complexes (occasionally termed ironstones or rødbergites) contain a slightly higher proportion of HREE, and much greater overall REE contents. These ironstones are often interpreted as the weathering products of the pristine calcite, dolomite, or ankerite carbonatites with the elevated HREE explained by preferential leaching of LREE during laterite formation. However, these ironstones are ubiquitous in carbonatites of all climate regions and often juxtaposed next to outcropping and perplexing unweathered carbonatite varieties. An alternative explanation is that the ironstones are the weathered products of highly fractionated ferrous HREE-rich plutonic alkali brine-melts ("natrocarbonatites"), which are commonly observed in experiments. Due to their high alkali contents they are not stable in the geological record and undergo rapid alteration. Therefore, these HREE-rich ironstones are not a marker of protracted weathering, but instead are the inevitable magmatic fractionation outcome of carbonatite melts. As such, they are expected to occur in all silica-poor carbonatite complexes. In silica-rich carbonatites, the ferrous alkali brine-melt is unattainable because it will react to form silicates such as phlogopite, aegirine, riebeckite, and arfvedsonite, preventing the formation of HREE-rich zones.

A New Century for Nickel; Assessing Global Reserves, Resources, and Sustainability of the Devil's Metal

Simon M. Jowitt[1], Gavin M. Mudd[2]
1. University of Nevada Las Vegas, Las Vegas, NV, USA, 2. RMIT University, Melbourne, VIC, Australia

Nickel is fundamentally important to modern life and society and is vital for the production of stainless steel, specialty alloys, electroplating, battery production, and more. Global nickel demand is expected to soar as the world transitions to a low- to zero-CO_2 and carbon neutral energy and transport future that increases battery use for energy storage associated with renewable energy systems. This means that understanding the current nickel market and known nickel resources and reserves are key to enabling the energy transition, including the environmental, social, and governmental (ESG) challenges that may prevent nickel project development or may hinder the continuation of current operations. This presentation provides a comprehensive global overview of reported Ni resources and reserves for 2018 classified by mineral deposit type. The majority of nickel has been and will continue to be sourced from laterites and magmatic sulfide systems; other deposits have been mined for nickel although production from these remains relatively small. Our database includes 653 nickel deposits with at least in-ground resources and often reserves, yielding 346.0 Mt of contained Ni in resources that, if all converted into resources and hence production, could meet ~153 years of Ni demand at 2018 global production rates (~2.25 Mt Ni). Comparison of these data to 2011 data indicate that magmatic sulfide resources are keeping pace with depletion by mining whereas laterite resources and reserves are lower, suggesting the latter can be more comprehensively assessed during early resource and reserve estimation. Known nickel resources and reserves are sufficient to continue current levels of production for several decades. However, the nickel mining sector faces a number of challenges that may change this, including increased demand from electric vehicles and batteries and potential supply restrictions relating to increased ESG challenges to the mining industry (including and geopolitical challenges like the Ukraine-Russia crisis).

Foliation Boudinage Structures at the Mount Isa Cu-Pb-Zn System, Australia

Benjamin J. Williams[1], Thomas Blenkinsop[1], Richard Lilly[2], Patricia Ila'ava[3]
1. Cardiff University, Cardiff, United Kingdom, 2. University of Adelaide, Adelaide, SA, Australia, 3. Queensland Metals Technical Services, Mount Isa, QLD, Australia

Foliation boudinage structures are vein-like boudin structures that form in highly anisotropic and homogenous rocks during layer parallel extension and layer normal shortening events. These structures may have a more significant role in the formations of ore bodies than previously thought. At Mount Isa, the finely laminated steeply West dipping Urquhart Shale contains the breccia hosted Cu and stratiform Pb-Zn ore bodies and provides the homogeneity and anisotropy required for the formation of foliation boudinage structures. Through detailed structural drill core logging in the Mount Isa Cu-Pb-Zn system, the necklines of drill-core scale foliation boudinage structures have been shown to plunge gently to moderately to the North or South and indicate the approximate intermediate strain directions during a regional ENE-WSW shortening event. These foliation boudinage structures have an initial rim of quartz and dolomite, with a later pyrrhotite fill with minor co-genetic to late chalcopyrite in some samples. Increasingly deformed and altered lithologies exist closer to the Cu ore bodies, where there is a general trend of increasing high-grade Cu with deformation. The abundance of pyrrhotite and the relative absence of Cu from the infill of the small-scale foliation boudinage structures suggests the structures are within the silica-dolomite and pyrrhotite mineralisation zones whilst being outside of the chalcopyrite breccia zone of high-grade Cu. Foliation boudinage structures are located predominantly in the least deformed shale lithologies, and drillhole intervals with foliation boudinage structures show an overall lower Cu% when compared to the entire drillhole length. The small-scale foliation boudinage structures could be used as a targeting and localisation tool for future exploration drilling at the Mount Isa system. Further work is needed to test for the existence of larger scale (metre to 100s metres) foliation boudinage structures which might have played an important role in ore body localisation and orientation.

Critical Minerals for Our Energy Future: Geology and Ore Deposit Models

Frances Wall
Camborne School of Mines, Met4Tech and HiTechAlkCarb projects, University of Exeter, Penryn, United Kingdon

New technologies, especially energy technologies require not only greater amounts of mined raw materials but also a wider range of elements than ever before. Our techno-world is made of clever and complex elemental mixtures. Some of these raw materials are 'critical' because they are economically important but, with only relatively small amounts needed, their supply is vulnerable to disruption. This is good news for geologists. We need to understand the concentration of critical minerals of elements like rare earths, lithium, cobalt and more, and this is an exciting opportunity for lots more geology. What is not such good news, is the projected exponentially increasing demand for minerals, driven by population growth, increasing wealth and the metal-intensive nature of clean energy technologies. Slowing this rate of increase whilst maintaining the energy transition is one of the most important challenges for humankind. As geologists we must integrate mitigation strategies into our new geomodels. The approach used in the EU H2020-funded HiTech AlkCarb project, which created new geomodels for rare earths and other critical raw materials in alkaline rocks and carbonatites, was to integrate deposit-related environmental and social aspects into our new conceptual mineral system geomodel. This included information on process mineralogy and radioactivity. In the NERC-funded LiFT project we are using a life cycle assessment (LCA) approach to compare the different types of lithium deposits and quantify impacts. Using LCA early in exploration leads to better mine and process design. Integrating LCA into resource models enables environmental as well as economic mine scheduling. In the UKRI Circular Economy Centre, Met4Tech, we are working on a geomodel for SW England that takes an integrated, regional view; combining Li,Sn,W ores, waste and by-products, legacy waste, and geothermal energy. We are investigating if this can help to better quantify and thus lower environmental impacts.

Rare Earth Element Mobilization and Enrichment in Alkaline Complexes

Dennis O. Achoki
RWTH Aachen University, Aachen, Germany

Rare Earth Elements (REEs; lanthanide series, Sc and Y) have vital applications in today's society and demand for them has been projected to rise in the next decades. Their most important applications include manufacturing of magnets used in motors of electric vehicles and wind turbines; in making components of smart phones and computers; and in a number of light-weighting applications for the aviation industry. Alkaline igneous complexes have emerged as fertile provinces for REEs and Y. Fractional crystallization of the parent alkaline magma leads to an enrichment of incompatible elements, of which REEs and Y comprise, in late-stage magmatic-hydrothermal fluids. The exposed roof zone of an alkaline intrusion and surrounding fenite aureole in the Iivaara alkaline complex of Finland provide a natural laboratory to study the origin, mobilization and enrichment of REEs and Y. Petrographic and textural examination of 10 hand samples and 15 thin sections representing all the rock types found in Iivaara reveal the complex to be comprised of ijolite, nepheline syenite, vein type pyroxenite, syenite fenite, quartz fenite and gneiss country rock. Apatite, a major host of REEs and Y, has been identified in all rock types of the Iivaara complex. Polarization microscopy and high resolution Back Scatter Electron (BSE) imaging show magmatic and hydrothermal textures for the apatites and related minerals. Hydrothermal fluid – apatite interaction is recorded as compositional zoning rims and dissolution features. Electron Probe Microanalysis (EPMA) and Laser Ablation-Inductively Coupled-Mass Spectrometry (LA-ICP-MS) have been conducted to analyse for the major and trace elements in apatite. REEs and Y have been detected in apatites of all rock types albeit with variability among the different rock types and within individual apatites. In this work I propose that REEs and Y mobilization and enrichment take place as element exchange and addition reactions that involve charge balancing.

Morphotextural and Chemical Analysis of Proximal Gold Placers and Epithermal Primary Sources: the Serra Dourada – Bananal Deposit (Aguapeí Belt, Brazil)

Kelvin d. Alves[1], Santos B. Sanchez[1], Juan G. Barreiro[1], Javier E. Villar[1], Jose Manuel C. Pietro[2]
1. Universidad de Salamanca, Salamanca, Spain, 2. Servicio de Difracción de Rayos-X, Universidad de Salamanca, Pza. de Los Caídos s/n, cp 37008, Salamanca, Spain

Understanding the evolution of gold geochemistry and morphology after its liberation from lode into alluvial systems, is critical to define efficient prospection strategies. Most morphological changes occur in the short range (transport distances <5km), but the chemical evolution is unclear. We have investigated those features in gold particles from the Bananal proximal placer (BP), and Serra Dourada (SD) epithermal deposit. Primary gold (SD) is formed by monocrystals and polycrystals with sizes between 1.3 and 11.75 mm, octahedra habit and, sometimes, hoppered growth. Placer gold (BP) nuggets range between 1.95 and 6.95 mm, and only 17% of them preserve clear primary morphologies (MG1). Most of placer nuggets (83%) exhibits mechanically reworked shapes (MG2), with mean CFI, CSF and Shilo indexes of 2.40, 0.46 and 1.41. The distribution of morphological populations is compatible with data from other proximal placers (< 5km transport range). Textural and geochemical results show primary crystals (SD) are homogeneous Au:Ag alloys (Au1 £0.1% wt% Ag), without mineral inclusions, and no development of supergene Au-rich rim. About 12% of placer nuggets are chemically and morphologically equivalent to primary crystals (Au1 £ 0.6% wt% Ag). In this group the formation of a compositional rim is connected to the morphological evolution during transport. However most of placer particles (>80%) has a distinct geochemical signature: ternary alloys (Au:Ag:Cu) in the core (Au1), and mineral inclusions (Quartz, Pyrite, Cobaltite-gersdorffite). Interestingly, this last group shows no evident morphological signature but general MG2 ones. Our data point to Serra Dourada -Bananal deposits are genetically and spatially connected, and represent the evolution of an epithermal HS system in which a primary geochemical zoning exists. Erosion level, and colluvial-to-alluvial sedimentary context, were responsible of the morpho-chemical mixture founded in the Bananal placer.

A Study of Philipsburg's Polymetallic Lode Deposits, Granite County, Montana

Celine M. Beaucamp, Christopher H. Gammons
Montana Technological University, Butte, MT, USA

From the late-1800s to the mid-1900s, the 10 km^2 mining district just east of Philipsburg, Montana, produced over $2.4x10^7$ oz Ag, $3.6x10^7$ kg Zn, $1x10^7$ kg Pb, $1.8x10^6$ kg Cu, $4.5x10^5$ short tons of battery-grade MnO_2, and significant W. Although its silver production was surpassed in Montana only by the famous porphyry-lode deposits of Butte, the Philipsburg district has never been studied with modern methods of ore-deposit research. This study aims to re-evaluate the district's geochemistry, structural history, and economic potential, including critical metals. Mineralization occurs as quartz + rhodochrosite + barite veins and replacements, hosted by Precambrian to Mesozoic metasedimentary rocks and the 75 Ma Philipsburg granodiorite stock. A small 65 Ma Mo-porphyry outcrops on the district's northern edge, but its relationship to polymetallic vein mineralization is unclear. Mineral zonation in the district consists of a low-sulfidation suite (arsenopyrite + tetrahedrite + Fe-rich sphalerite) at the periphery, grading to a higher sulfidation suite (enargite + tennantite + Fe-poor sphalerite) in the center. Stable S-isotopes are lighter in the central area (near 0 per mil) than in the rest of the district (+2 to +6 per mil). Sphalerite from the central area fluoresces unusually bright colors under longwave UV (365 nm) light. LA-ICP-MS analysis reveals that the red fluorescence is correlated to a very low Fe-content, coupled with high concentrations of Cu, Ge, Ga, In, W, and Ag in the sphalerite lattice. Concentrations of Ga and W are particularly high (up to 4997 and 2834 ppm, respectively) and may be of economic importance. The Philipsburg district has many geological and geochemical similarities to the nearby Butte porphyry-lode district and likely belongs to the same clan of Cordilleran-style, zoned, polymetallic-lode deposits. Re-examination of historic mining districts such as Philipsburg may help address the rising need for domestic production of rare metals.

Structural Control for the Hosting of Gladys Belt in the Huachocolpa Mining District

Ivan E. Branes, Miguel Miranda, Mario Rosas, Omar Yaringano, Jim Torres
KOLPA MINING COMPANY, LIMA, Peru

The Gladys project is located 6 km to the SW of the Huachocolpa district, Huancavelica, Peru. It is a system of veins to the East of Veta Bienaventurada in current operation by Compañía Minera Kolpa.

Bienaventurada vein is the master vein with strike N60°E and Dip to the SE has a longitudinal extension of 4 km and its tensionals (E-W) outcrop at surface.

The Bienaventurada vein performs dextral movement with normal dip and tensional forms to the West, the Tatiana (E-W), Mariteresa and Marisabel veins have been identified, towards the East side only the Elizabeth and Jessica veins were identified, which are parallel to Bienaventurada. However, a vein with strike E-W on the east side of Bienaventurada was not recognized. This target was called Gladys belt.

The objective of the 2021 drilling program was to find more tensional veins to the east of Veta Bienaventurada, these 3 structures (Sofia, Techita and Stephanie) being called the Gladys belt. Sofia parallel to Bienaventurada vein, Techita parallel and antithetical to Bienaventurada vein. Stephanie with strike E-W and antithetical to Bienaventurada vein.

7 holes were drilled, totaling 5,260 meters of drilling, only 3 holes intercept the Stephanie vein. Significant Ag results were obtained in drillhole DDH-16-21 corresponding to the Stephanie vein @2.15 m. 19.27 Ag Oz/MT, 0.72% Cu, 13.4% Pb and 5% Zn.

Currently Sofia, Techita and Stephanie veins have entered into production and a drilling program is development to find the ore shoots of Gladys belt.

Understanding the tectonic concept of the Bienaventurada vein was decisive for the discovery of the Gladys belt.

Fresh Interpretation of PGE-Ni-Cu-Coo Mineralisation in the Northern Limb of the Bushveld Complex

Kate R. Canham[1], David A. Holwell[1], Andy Lloyd[2]
1. University of Leicester, Leicester, United Kingdom, 2. Anglo American South Africa, Johannesburg, South Africa

The Northern Limb of the Bushveld Complex, South Africa, is host to one of the Earth's largest resources of platinum-group elements (PGE), along with significant Ni, Cu and Co. The limb is best known for the world class Platreef deposit. Located in Critical Zone stratigraphy, the Platreef is a complex marginal style PGE-Cu-Ni-Co deposit, comprised of a series of pyroxenite sills resting upon variable Archean-Paleoproterozoic country rocks. Whilst substantial research has been undertaken on the near surface Platreef resources and its variability along strike, it is now possible to study the deeper Critical Zone. As mining and exploration activities advance to deeper levels, it is becoming evident mineralisation in the Critical Zone is not only restricted to the PGE rich Platreef that has been mined at surface. Base metal sulfide (BMS) enriched units, chrome units and multiple PGE units have been observed in the deeper Critical Zone, with deeper Critical Zone mineralisation flattening out and thickening in places, transposing into a more stratiform reef-style mineralisation with significant PGE grade.

It is well established that both magmatic and fluid processes have played a part in developing the highly variable mineralisation along strike of the Platreef, however these processes are poorly understood in the deeper Critical Zone's mineralisation history, as is the observed layering of PGE, BMS and chrome units. This ongoing study integrates mineralogical studies, PGM and sulfide studies, bulk rock geochemistry and S isotopes to further unravel the nature and genesis of what is one of the world's greatest ore deposits. Additionally, this will aid in assessing the different processes, magmatic (fractionation), contamination and hydrothermal, which are responsible for the variability of mineralisation and metal budget within the Critical Zone.

Geological and Geochemical Characterization of the Chifumbazi Gold Prospect, Irumide Belt, Tete Province, Northern Mozambique

Euclesia P. Cossa[1, 4], Andrea Agangi[1], Ryohei Takahashi[1], Pearlyn Manalo[1], Akira Imai[2], Vicente Manjate[3]
1. Akita University, Akita, Japan, 2. Kyushu University, Fukuoka, Japan, 3. National Institute of Mines, Maputo, Mozambique, 4. Instituto Superior Politecnico de Tete (ISPT), Tete, Mozambique

The Chifumbazi gold prospect is located in the northwestern part of Tete province, Mozambique. The prospect lies on a northern terrane of Tete Province which is underlain by supra-crustal volcano-sedimentary rocks of the Mualázi Group in the Southern Irumide Belt. Although Mozambique has been geologically mapped on a 1:250 000 scale, gold mineralization studies are lacking in Tete Province. This study aims to describe the gold mineralization and associated hydrothermal alteration in the Chifumbazi prospect, based on petrography, whole-rock geochemistry, and sulfur isotope data.

The mineralization, in the Chifumbazi prospect, occurs as carbonate-quartz-chlorite, quartz, and chlorite veins and veinlets and, sometimes, disseminated in gneiss and meta-granodiorite host rocks. However, the mineralized veins and veinlets occur along with the discordant fracture system within a Pan African shear zone. The meta-granodiorite consists mainly of plagioclase, biotite, hornblende, and quartz, with secondary calcite, dolomite, sericite, and chlorite; while the gneiss is composed of quartz, biotite, and opaque minerals. Hydrothermal calcite, chlorite, and sericite occur either as small veins and veinlets or as replacing primary minerals. Furthermore, the meta-granodiorite shows enrichment in LREE, depletion in HREE, and a negative Eu anomaly.

The ore minerals in the Chifumbazi prospect consist mainly of pyrite, pyrrhotite, chalcopyrite, and minor covellite, galena, magnetite, and bornite. Visible gold was not observed, however, gold contents up to 5 ppm were detected by inductively coupled plasma mass spectrometry. Thus, we suggest that gold in the Chifumbazi prospect occurs as a solid solution in sulfides.

The $\delta^{34}S_{CDT}$ of pyrite and chalcopyrite varies in a narrow range, between +1.1 and +2.9 ‰, suggesting a restricted source of sulfur, which could be magmatic or metamorphic in origin.

Based on geological setting, mineralogy, alterations, and sulfur isotope signatures, the Chifumbazi gold mineralization can be classified as an orogenic-type gold deposit.

The Nazca-Ocoña Metallogenetic Belt, Arequipa, Peru: Geological Characteristics of Quartz-calcite-sulfide Veins

Jorge Crespo[1], Elizabeth Holley[1], Katharina Pfaff[2], Madeleine Guillen[3], Sergio Ticona[3], Isaac Simon[2]
1. Colorado School of Mines, Department of Mining Engineering, Golden, CO, USA, 2. Colorado School of Mines, Department of Geology and Geological Engineering, Golden, CO, USA, 3. Universidad Nacional de San Agustin de Arequipa, Departamento de Geologia y Geofisica, Arequipa, Peru

The Nazca-Ocoña metallogenic belt in southern Perú is host to more than 70 recognized Au-Ag and base metal deposits. These deposits are spatially associated with Cretaceous intrusive rocks of the coastal Batholith in a zone that extends more than 350 km along the Central to Southern part of Perú. In order to establish a revised genetic model for precious metal bearing veins in the Nazca-Ocoña belt, we conducted detailed studies at two representative deposits in this metallogenic belt: Mollehuaca and San Juan de Chorunga. At both sites, small-scale mining has focused on quartz-calcite-sulfide veins hosted in granodiorite. Here we present the results from geological field mapping, field-emission scanning electron microscopy, electron microprobe analysis, nano-scale secondary ion mass spectrometry, laser ablation inductively coupled plasma mass spectrometry (LA-ICP-MS) on pyrite, and fluid inclusion petrography of quartz to elucidate the ore forming processes.

Ore mineralogy is characterized by pyrite, chalcopyrite, arsenopyrite, galena, sphalerite, native gold, Au-tellurides, and electrum. Native gold occurs disseminated throughout the veins as micro to sub-micron inclusions in pyrite, and electrum occurs in pyrite as cavity and fracture fill. The LA-ICP-MS analyses show that pyrite has complex internal zonation and are host to precious metals (Au, Ag), base metals (Co, Ni, Cu), and metalloids (As, Sb) as lattice-bound impurities and sub-micron inclusions. San Juan de Chorunga pyrite has $\delta^{34}S$ values ranging from 0.4 to 1.4 ‰, and Mollehuaca pyrite $\delta^{34}S$ values range from 5.8 to 7.4 ‰, representing a magmatic-hydrothermal sulfur. Fluid inclusion petrography of these veins show recrystallized quartz with implosion and explosion fluid inclusion textures, indicative of pressure variations during quartz vein formation. Our observations suggest that the majority of the gold in these veins formed in association with a shallower hydrothermal event, potentially in association with the emplacement of shallow intrusions in the Late Cretaceous.

Sediment-hosted Gold Mineralization in the Gegalaw Deposit, Central Myanmar

June B. Born, Yasushi Watanabe, Patthana Bounliyong
Akita University, Akita, Japan

The Gegalaw is one of several gold deposits situated along the Sagaing Fault, which marks the boundary between the Central Myanmar Basin in the west and the Mogok metamorphic belt in the east. Gegalaw is hosted by Triassic calcareous sedimentary rocks, which are unconformably overlain by sandstone of the Tertiary Male Formation. The host calcareous shale contains quartz, kaolinite, chlorite, illite, and carbonaceous materials, with minor amounts of framboidal and aggregated pyrites. Ore-grade rocks (up to 2.4 g/t Au) are found in the breccia and strongly silicified zones. The ores are mainly composed of quartz, kaolinite, chlorite, illite, calcite, dolomite, ankerite, and sulfides (mainly pyrite). Based upon the textural and chemical composition classified pyrite into three types: 1) pyrite with porous core and compact rim zoning texture; 2) arsenopyrite pseudomorph; and 3) well-crystalline homogenous pyrite. EPMA identified the highest gold content up to 1080 ppm in the compact rim part of zoned pyrite, together with high As content up to 7 wt%. $\delta^{34}S$ value of pyrite from both host and mineralized rocks yielded a range of -18.9 to -0.5‰, and the $\delta^{34}S$ of the framboidal pyrite in the host rock (-17.3‰) is similar to the $\delta^{34}S$ of pyrite of the highest Au-grade rock sample (-18.9‰). Suggesting that auriferous pyrite sulfur is sedimentary in origin. Nevertheless, the sulfur isotopic composition of two pyrite samples from the mineralized rock yielded heavier $\delta^{34}S$ values (-4.9‰ and -0.5‰), which suggests a contribution hydrothermal fluid source. Illite in the vein of mineralized rock yielded a K-Ar age of 40.26±0.91 Ma. The research results indicate the Gegalaw gold mineralization occurred infiltration of hydrothermal fluid into the host rock. Furthermore, the age of illite suggests a link to the early stage of the Sagaing Fault activity that was related to the collision between Indian and Asian Plates.

SEG 2022 Conference: Minerals For Our Future

Rare Earth Elements (REE) Potential in the Cornudas Mountains, Southern New Mexico

Virginia T. McLemore, Nels Iverson, Mason Woodard, Haley Dietz, Ethan Haft, Snir Attia, Richard Kelley
New Mexico Bureau of Geology and Mineral Resources, New Mexico Institute of Mining and Technology, Socorro, NM, USA

Re-examination of the rare earth elements (REE) deposits in the Cornudas Mountains is warranted in light of today's economic importance of critical minerals, including REE that are essential in most of our electronic devices. New mapping and petrographic, $^{40}Ar/^{39}Ar$ geochronology, and geochemical analyses are underway. The Cornudas Mountains form the northern Trans–Pecos alkaline magmatic province in the southern part of the North American Cordilleran alkaline-igneous belt. The igneous rocks in the area were emplaced ~36-25 Ma, just prior to or during the early phases of Rio Grande rift extension, and consist of 1) larger nepheline syenite–syenite laccoliths and plugs, 2) phonolite plugs, sills, and dikes, and 3) smaller syenite plugs and dikes that intrude Permian and Cretaceous sedimentary rocks. New USGS geophysical data indicate that some of these intrusions extend deep into the subsurface, with additional intrusions potentially buried in the subsurface. The focus of REE exploration is along the outer edge of the Wind Mountain nepheline syenite laccolith, as well as within syenite–phonolite dikes, plugs and altered areas in Chess Draw. Some samples contain as much as 3110 ppm total REE, hosted in REE-bearing minerals (eudialyte, monazite, bastnäsite, calciocatapleiite). We incorporate whole rock and clinopyroxene chemistry of each intrusion into the clinopyroxene-liquid geothermobarometer (Putirka, 2008) to determine the temperatures and pressures of emplacement. This thermometer provides higher crystallization temperature estimates for the syenite intrusions (821-1019°C) than the phonolite sills (763-797°C). We then use the barometric estimates (1.4-2.9 kbar) to calculate emplacement depths (5.2-10.8 km). Pairing these depths with the new geochronology, we can estimate minimum exhumation rates for intrusions in the Cornudas Mountains that range from 0.1-0.4 mm/yr, with faster exhumation rates for younger intrusions. Estimating crystallization temperatures and exhumation rates provide additional information to aid in developing a model for the formation of REE deposits.

Characterization of Gold from the Liberty Area, Washington

Timothy A. Miller, Angela Halfpenny
Central Washington University, Ellensburg, WA, USA

Gold (Au) is a metallic element and mineral that has been used for thousands of years and is one of the foremost mediums of exchange in most countries. Societal demands for critical minerals and vital metals, like gold, supports the need for continued research into these systems. The chemistry and crystallography of gold can be used to interpret its provenance and genesis. This research will focus upon understanding the fundamental processes that control the concentration, deposition and growth of gold in the Liberty area, Washington. The area presents a unique field site for studying the controls on gold mineralization, preservation, and weathering, as it displays a variety of gold textures including nuggets, placer grains, wire gold and crystalline lode gold. This research will work on the 20-acre Gold Reserve (GR) mining claim located directly south-east of the Liberty townsite. Areas that contain gold will be identified using a metal detector and detailed mapping performed to record the mineralization structures, local faulting, contacts and other geological features. A non-destructive, Bruker Tracer 5i, portable X-ray fluorescence spectrometer (pXRF) will be used to measure the elemental signature to identify areas containing gold within quartz/carbonate mineralization of the GR claim for sampling. It is unclear how much of the placer gold found in the Liberty area is sourced from the local primary crystalline gold found in hard rock mineralization. A comparison between the two can be made by using scanning electron microscopy (SEM) to characterize the chemistry and crystallography, to determine if the placer deposits could have been sourced from the local in-situ gold. The results will improve our understanding of the paragenesis and overall evolution of the gold within the Liberty area. This knowledge can be applied to other key commodities required to fuel our global society.

Mineralogical Evolution and Gold-Silver Zonation in the Segovia-Remedios Mining District, Colombia

Ricardo S. Molano[1], Juan C. Molano[1], Ariel Cadena[1], Alessandro Cecchi[2], Julian Ceballos[2], Leonardo Santacruz[2]
1. Universidad Nacional de Colombia, Geociencias, Bogotá, Colombia, 2. Gran Colombia Gold Corp, Medellín, Colombia

The Segovia-Remedios Mining District (SRMD) is located on the eastern flank of the Central Cordillera of Colombia. It is considered the major gold producer in the country. In 2020, Gran Colombia Gold Company (GCG) extracted around 220.194 gold ounces, representing 13.1% of the gold obtained in Colombia that year. The mineralization comprises quartz veins hosted in granodiorite Jurassic rocks, exhibiting NE-SW and NW-SE orientations. Geochemical data reveals variations in Au:Ag ratios: 1:1 and 1:400 in the NW and SE sectors of the district, respectively. However,we still don't know ocurrence of silver mineral phases and the metallogenic processes associated with gold-silver zonation.

Ore petrography descriptions and EPMA analysis in samples collected in four mines along SRMD reveal *three hydrothermal stages. Stage I* is characterized by massive milky quartz + subhedral pyrite, which is later affected by shear episodes related to Stage II formation. *Stage II*consists of pseudomorphic and early pyrite (after marcasite-pyrrhotite) fractured by the zonal sphalerite (2.58 – 17.61 molar % FeS) + a later galena event. Besides at *Stage II*some late euhedral pyrite occurs. Finally, *Stage III* is identified by the occurrence of carbonate sheeted veinlets. Gold mineralization occurs as electrum associated with *Stage II* in three different styles. At the northwest area, electrum appears as small inclusions of zoned sphalerite; also related to galena in fractured pseudomorphic pyrite and as microscopic inclusions in euhedral pyrite. Nevertheless, on the SE, electrum can only be found associated with galena, as well with tetrahedrite, freibergite, pyrargyrite, and argentite.

Sulfide paragenetic variations and silver sulphosalts ocurrence suggest changes in the sulfidation state as the first cause of gold-silver zonation. In addition, we are conducting micro-thermometric analysis in quartz and sphalerite of *Stage II* to know fluid facts and processes associated with silver-gold precipitation mechanisms and their zonations.

The Metallogenesis of the Orogenic Gold Fields in the Karagwe-Ankole Belt in Central Africa:" a Systematic Review of the Current Understanding"

Rub'son H. N'nahano[1], Huan Li[1], Moise Luemba[2], Claude Nambaje[3]
1. Central South University, School of Geosciences and Info-Physics, Department of Geological Engineering and Geological Resources, Changsha, Hunan, China, 2. China University of Petroleum,School of Geosciences,Changjiang West Road, 66 Hungdao District, Qingdao, Shandong, 266580, China, Qingdao, China, 3. Indian Institute of Science, Centre for Earth Sciences, Bangalore 560012, Bangalore, BR, India

The gold metallogenic evolution of Central Africa has not been fully investigated in detail despite its huge reserves and a long mining history that has gone over a century. Most of the gold deposits of the main Kibara belt, known as the Karagwe Ankole Belt in some literature, are located in the eastern part of the DR Congo, named the Twangiza-Namoya Gold Belt (TNGB). These orogenic gold deposits are found in Proterozoic, Mesoproterozoic, and Neoproterozoic geological terranes. The variability of ore fluid genesis with regard to the complexity of the geological formations records makes the Central Africa's magmatic-hydrothermal systems challenging compared to the other orogenic gold world fields. Adding to that, the lack of knowledge of the particular hydrothermal processes that were involved in carrying and depositing the gold mineralization during the entire Precambrian era is making the region scientifically undiscovered in the field of economic geology. This work reviewed the systematic fluid inclusions, sulphur and boron isotope studies done in the region with the purpose of getting the preconception of the gold metallogenic model of the Karagwe-Ankole Belt (KAB). The gold deposits in the KAB are being supplied by fluids derived from sedimentary formations. Although some fluids are proved to be of magmatic and/or metamorphic origin, the most influential fluids are derived from sedimentary origins occurred during the greenschist facies metamorphism. Two types of gold bearing fluids have been constrained: the gold bearing fluid derived from the crystallization of the regional G4 specialized granites associated with the endowment of Nb-Ta-Sn mineralization and the ore fluids associated with the main fluid activity occurring during the early Pan-African orogen. This article concludes on the future research investigation that has to be done regarding the H-O-C-S and Cl systematic isotopes of each single deposit in the Karagwe Ankole Belt.

Distinctive Chemical Composition of Gold-Ore-Forming Magmas

Carolina Nogueira Mafra, Robert R. Loucks, Marco Fiorentini
Centre for Exploration Targeting, University of Western Australia, Perth, WA, Australia

Magmatic-hydrothermal gold deposits generally form in convergent margins or post-subduction settings and tend to cluster in restricted regions of the crust (as a metallogenic province). Magmatic fertility is affected by geochemical inheritance and metal budget derived from the melt source as well as by a chain of processes that occur during magma emplacement. The contribution of a metasomatized subcontinental lithospheric mantle (SCLM) source has been regarded as the most preeminent parameter for gold fertility. Geodynamic settings that favor the participation of these pre-enriched portions of the SCLM in the melting column and magma chemistry also have petrogenetic control on the gold endowment of magmatic-hydrothermal systems.

The contrast in the melt source region of gold-fertile and gold-infertile magmas is emphasized by the differences in chemical composition between primitive melts spatially and temporally associated with ore-forming magmas in major gold districts and primitive magmas from magmatic belts unrelated to gold mineralization. The fertile ones are enriched in highly incompatible elements compared to infertile ones, whereas they have a comparable concentration of mildly incompatible elements. In order to investigate the participation of an enriched source in the melting column of ore-forming magmas, the chemical composition of a large dataset of ore-related magmas from gold-rich deposits is compared with the chemical composition of intrusive rocks from magmatic belts that lack gold mineralization.

Low- to high-K_2O calc-alkalic magmas that exsolve a gold-ore-forming magmatic-hydrothermal fluid can be distinguished from infertile magmas by a series of trace element ratios having a highly incompatible element in the numerator and a mildly incompatible element in the denominator. Although this trend is evident for calc-alkalic magmas, potassic alkalic magmas are often enriched in those highly incompatible elements regardless of their metallogenic signature and the enigma connecting alkaline magmatism and gold enrichment still needs investigation.

New Investigations of the REE-Nb Carbonatite Deposits of Southern Ravalli County, Montana, USA

Sarah Risedorf[1], Christopher H. Gammons[1], Peter Mejstrick[2]
1. Montana Tech, Butte, MT, USA, 2. U.S. Critical Materials, Salt Lake City, UT, USA

With funding from U.S. Critical Materials and the U.S. Army Research Lab, Montana Tech has begun a comprehensive investigation of the REE-Nb carbonatite dikes of southern Ravalli County, Montana. The district contains a large number (> 40) of thin (< 4 m), discontinuous carbonate bodies hosted by deformed and metamorphosed meta-igneous rocks (amphibolite, meta-gabbro, felsic augen gneiss) previously dated to circa 1.4 Ga. The meta-igneous complex and associated REE-rich carbonatite bodies extend into eastern Idaho, and are on strike with the Lemhi Pass Th-REE district, circa 100 km to the southeast. The meta-igneous complex is bordered to the north by relatively unmetamorphosed siliclastic sedimentary rocks of the Belt and/or Lemhi Supergroups; however, no carbonatite bodies are known to exist in the meta-sediments. The meta-igneous complex may represent a flare-up associated with the onset of Mesoproterozoic rifting. Geochemistry of channel samples (n = 40) shows a possible evolution from early magnesio-carbonatite (dolomite + apatite) to later calcio-carbonatite (calcite + barite + Mg-riebeckite), with enrichment in light REE (up to > 10 wt% ΣREE) and Nb (up to 0.2% Nb) in the form of monazite, ancylite, allanite, columbite, and a host of minor phases. Sparse sulfides include pyrite, pyrrhotite, chalcopyrite, Co-Ni-As sulfides, and molybdenite, the latter of which has been submitted for Re-Os dating. The surrounding mafic country rock is locally altered to coarse phlogopite ("glimmerite"). Carbonate isotopes are consistent with a magmatic origin for the dolomite-rich carbonatites, and possible involvement of high-temperature meteoric waters for the more mineralized calcio-carbonatites. More C-O-H-S stable isotope work is in progress, as well as detailed mineralogy, trace element geochemistry, field mapping, and geochronology. Other partners in this project include the U.S. Geological Survey, the Montana Bureau of Mines and Geology, and the Idaho Geological Survey.

Exploration Model for Base and Precious Metals in the Eastern Gavião Block, São Francisco Craton, NE Brazil: Geochemical and Tectonic Constraints

Ricardo R. Spreafico, Antônio Marcos V. de Moraes, Michele C. Santos, Fabiano C. Livramento
Mining Company of the State of Bahia, Salvador, BA, Brazil

The whole-rock geochemical study of Neoarchean volcanic belts, emplaced in Paleoproterozoic large-scale tectonic lineaments, can provide models for exploration programs in the eastern Gavião Block (São Francisco Craton, Brazil), which is bordered by the north-south trend Contendas-Jacobina lineament (CJL), along with approximately 200 km. The Mundo Novo greenstone belt (MNGB) is hosted in the northern CJL and is composed in the middle unit by metabasalts that feature a vector from the MORB-OIB array to the volcanic arc array in the Nb/Yb-Th/Yb diagram, with a pattern from Calc-Alkalic Basalts (CAB) to Island Arc Tholeiitic basalts and N-MORB in the ternary diagram of the Th-Hf-Nb system. Low Zr values and Zr/Y and Ta/Yb ratios indicate an oceanic arc provenance for the MNGB, based on discrimination diagrams of Zr-Zr/Y and Ta/Yb-Th/Yb. The Contendas-Mirante belt (CMB) is hosted in the southern CJL and is composed in the lower unit by metabasalts and basaltic metandesites that are grouped in the volcanic arc array in the Nb/Yb-Th/Yb diagram and do not overlap the MORB-OIB array. These rocks are strictly plotted in the CAB field in the Th-Hf-Nb ternary diagram. High Ta/Yb values also suggest an active continental margin provenance for the CMB, based on Ta/Yb-Th/Yb diagram. The coexistence of oceanic and continental volcanic belts in distinct parts of the CJL indicates different orogenic patterns. The oceanic provenance of the MNGB suggests an accretionary event was predominant in the northern CJL, although a collisional event occurred in the southern CJL, in which the CMB would represent a continental margin setting, with the oceanic crust completely consumed. These different settings are important guides for base and precious metals exploration in the eastern Gavião Block, indicating important potential for gold in the CMB and southern CJL, and the main potential for base metals in the MNGB and northern CJL.

Depositional History and Gold Potential of the Ament Bay Assemblage in the Sturgeon Lake Greenstone Belt, Northwestern Ontario, Canada

Michael Tamosauskas[1], Ross Sherlock[1], Chong Ma[1], Rasmus Haugaard[1], Robert Lodge[2]
1. Laurentian University, Sudbury, ON, Canada, 2. University of Wisconsin-Eau Claire, Eau Claire, WI, USA

Evaluating the sedimentology, provenance and stratigraphy of the Ament Bay assemblage can provide insights into the geologic history of the Sturgeon Lake greenstone belt and the metallogeny of Archean extensional basins. The Sturgeon Lake greenstone belt makes up the easternmost portion of the western Wabigoon terrane of the Superior craton, and is comprised of mostly Neoarchean volcanic assemblages, minor siliciclastic successions, and calc-alkalic to alkalic intrusions. The Ament Bay assemblage is the youngest supracrustal assemblage of the Sturgeon Lake greenstone belt and dominantly consists of polymictic conglomerates, subarkosic to arkosic arenites and greywacke-mudstone sequences. These lithofacies are intruded by alkalic porphyritic rocks, which are also incorporated as clasts in Ament Bay assemblage conglomerates, indicating a coeval relationship between sedimentation and magmatic activity. Features such as abundant planar bedding, spatially associated faults and locally derived debris suggest that Ament Bay assemblage sedimentation occurred in a subaqueous setting within a transtensional pull-apart basin. The lithofacies of the Ament Bay assemblage resemble the auriferous 2670-2680 Ma Timiskaming assemblage, a prominent regional exploration target for gold in the Abitibi greenstone belt. In the Timiskaming basin, alkalic magmatism and coeval deep-reaching extensional faults are critical factors for gold metallogeny. Because these mechanisms are postulated to have occurred during the formation of the Ament Bay assemblage, this study provides insights into its gold potential and considers possible factors which may have impeded metal endowment.

Underrated and Overlooked: the Magmatic–hydrothermal Transition Recorded by Trace Elements-in-Quartz

Leonidas C. Vonopartis[1], Paul A. Nex[1], Judith A. Kinnaird[1], Laurence J. Robb[2, 1]
1. University of the Witwatersrand, Johannesburg, South Africa, 2. University of Oxford, Oxford, United Kingdom

In exploration, quartz is commonly considered an inconsequential mineral although, with the advent of analytical techniques such as LA-ICP-MS, this is changing. The ability to measure trace element variations in quartz provides a new window into the evolution of mineral deposits.

Granites are typically host to late-stage magmatic-hydrothermal mineralisation, such as Sn and other commodities that are considered critical metals. The Nebo, Bobbejaankop and Lease granites in the Zaaiplaats Tin Field of the Bushveld Complex, are host to various expressions of endogranitic Sn-mineralisation. Zaaiplaats comprises the Nebo Granite at the base that grades into the Bobbejaankop Granite and then the Lease Granite, which forms the fine-grained cupola. These granites exhibit an increasing degree of hydrothermal alteration respectively, with mineralisation restricted to the Bobbejaankop and Lease phases. High-grade Sn-bearing tourmaline-quartz hydrothermal pipes radiate up through the Bobbejaankop into the Lease Granite and terminate below the roof contact.

Trace element analysis of quartz from these granites exhibit evidence that supports the suggested fractionation and fluid-saturation models of ore genesis. The Al/Ti and Ge/Ti ratios in quartz increase from the base to roof and illustrate the sequential evolution of the system and an increasing degree of fluid-rock interaction. The quartz from the high-grade pipes is distinct with very low-Ti, high-Al and high-Ge, which indicates crystallisation from a very low-temperature late-stage and highly acidic fluid.

The trace element data displays a shift from an ordered magmatic fractionation-controlled evolution to a hydrothermally-controlled system that reflects more erratic trends. Therefore, trace element variations in quartz are able to define the point of fluid-saturation and record the magmatic-hydrothermal transition. The identification of the most evolved and fluid-saturated facies can guide exploration by indicating lithologies with the best mineralisation potential. The use of trace elements-in-quartz extends beyond granite-hosted deposits and may be applicable to various mineralised systems.

Textural Characteristics of Adularia Dendrites in Banded Quartz Veins from the Midas Deposit, Nevada

Lauren R. Zeeck[1], Thomas Monecke[1], T. J. Reynolds[2], Katharina Pfaff[1]
1. Center for Mineral Resources Science, Department of Geology and Geological Engineering, Colorado School of Mines, Golden, CO, USA, 2. FLUID INC., 1401 Wewatta St. #PH3, Denver, CO, USA

The middle Miocene low-sulfidation epithermal Au-Ag Midas deposit is located in the Northern Nevada Rift in north central Nevada. Since 1998, Midas has produced over 2.2 million ounces of gold and 26.2 million ounces of silver. Total measured and indicated resources at the end of 2021 graded 0.43 ounces/ton gold and 5.7 ounces/ton silver, though grades can locally exceed 10 ounces/ton. High-grade ores are confined to crustiform veins composed of quartz and adularia.

In this study, the adularia textures of high-grade ore samples were examined to identify the processes responsible for the formation of this gangue mineral. Micro-X-ray fluorescence elemental mapping revealed that the adularia commonly forms dendritic aggregates in the crustiform veins. The adularia dendrites can be up to several millimeters in size and are comprised of sub-millimeter, fine-grained rhombic crystals. Correlative microscopy involving optical microscopy, quantitative automated mineralogy, and scanning electron microscopy were employed to determine the relationship between adularia and ore minerals. Adularia is typically located in colloform silica bands that host only few ore minerals. Most of the ore minerals are hosted by distinct ginguro bands.

Phase separation of the hydrothermal liquids is widely interpreted to be the process responsible for the deposition of adularia and the ore minerals. Though adularia-bearing colloform bands in the crustiform veins at Midas contain ore minerals, these bands are not the primary host to ore. Adularia and the ore minerals have not been coprecipitated under the same conditions. It is hypothesized here that although adularia and the ore minerals may have formed by similar processes, the amount of vapor produced during phase separation may have controlled mineral deposition at any given time and location along the controlling structure.

Study and Discovery of Gold Bearing Conglomerates in the North China Craton

Baojin Zhao[1, 2], Zhenqun Xiang[3], Huaikun Li[3], Huichu Wang[3], Songnian Lu[3], Hang Chu[3], Guojie Hao[3]
1. Department of Environmental Sciences, University of South Africa, Roodepoort, South Africa, 2. School of Geoscience and Technology, Overseas Expertise Center for Deep Marine Shale Gas Efficient Development Innovation (111Center), Southwest Petroleum University, Chengdu, China, 3. Tianjin Center, China Geological Survey, Tianjin, China

Conglomerate types of gold deposits are predominantly hosted within cratons from the Achaean to the Paleoproterozoic in the world. The North China Craton is one of the three, including Yangtz and Tarim in China, which has significant exposure of the Proterozoic conglomerate horizons in the sedimentary basins of Shandong, Inner Mongolia, Shanxi, Hebei, Tianjin, Henan and even Jilin in the north and northeastern parts of China. The overlying area of the whole sedimentary basin is estimated to be larger than 100,000 km^2, which is similar or even larger in size than the Witwatersrand Basin in South Africa of about 80,000 km^2 and Tarkwa in Ghana of about 90,000 km^2. It is named the North China Basin by us for descriptive purposes. This paper presents the results of an initial field investigation, as well as mineralogical, geochemical and isotopic studies, through which a number of conglomerate horizons were identified in some subbasins with the economic potential of gold mineralization in the North China Craton. The gold-bearing conglomerates in the North China Basin are characterized by: 1) Brownish color of the matrix; 2) Gravels of single quartz and quartzite; 3) Gold-hosting minerals of hematite (>90%) and magnetite without sulfides; 4) Gold grades mostly below 1g/t with nuggets through mineral separation from the samples of outcrops; 5) Its formation in the Paleoproterozoic, such as Henan Songshan (about 2.2 billion years) and Tianjin Jixian (about 1.65 billion years); and 6) Gold mineralization similar to the Tarkwan Basin in Ghana, formed at about 1.9 Ga, rather than the Witwatersrand Basin formed at between 2.7 and 2.9 Ga. It is concluded that there is great potential for gold mineralization in the North China Basin.

The Milingui Iron Ore Deposit in Tchibanga, Gabon

Baojin Zhao[1, 2, 3], Gabriel Kamqa[1]
1. Havilah Consolidated Resources (Pty) Ltd., Johannesburg, South Africa, 2. Department of Environmental Sciences, University of South Africa, Florida, South Africa, 3. School of Geoscience and Technology, Overseas Expertise Center for Deep Marine Shale Gas Efficient Development Innovation (111Center), Southwest Petroleum University, Chengdu, China

The Milingui iron ore deposit is located some 80 km from the Mayumba deep sea Port, southwest of Tchibanga in Nyanga Province, which mineralisation consists of magnetite and hematite along approximately 20km of strike length and 1km width forming part of the Mayombe ridge. It occurs over two magnetic blocks (I and II) within the lease area totalling 996 km^2. Most historic exploration activities were focused on Block I which consists of 4 mineralized zones (A, B, C and D). The exploration work. carried out by BRGM (1952-1959), included pitting, trenching, Reverse Circulation (RC) and diamond drilling, resources evaluation and port development planning and BHP-Billiton (2007-2009) of airborne geophysical survey with limited grab-sampling.

Our recent work consisted of identification, re-mapping, sampling and geochemical analysis of the old pits and trenches, grab-sampling of the orebody outcrops, geophysical reinterpretation, metallurgical testing and feasibility studies. The main Fe mineralization types included laminated hematite (average 64% Fe), canga (58% Fe), itabirite (49% Fe) being developed on top of primary magnetite orebody (45% Fe) at the bottom. The current iron ore reserve is 178 Mt of ordinary iron ore and 20 Mt of high-grade direct shipping ore (DSO) (61-67% Fe) with an upside potential for economically beneficial Fe ore resources (>500 Mt).

The advantages of the deposits are: 1) Only 80 km to the planned Mayumba deep sea port); 2) No existing community and relocation issues; 3) Good exposure of orebodies on the mountain ranges; 4) Enough water supply for mineral processing in the area; 5) Enrichment of oil and gas supply for energy; and 6) Little issue of acid mine drainage due to the shortage of sulphides. Furthermore, there are other minerals such as marble and aggregates which can be co-developed in the lease and its surrounding areas.

Rionegro – a New Gold District in Colombia Including a Potential Witwatersrand and Carlin Type of Mineralization

Ricardo A. Valls Alvarez[1], Jorge C. Martin[2], Vadim Galkine[2]
1. Valls Geoconsultant, Toronto, ON, Canada, 2. Alicanto Mining Corp., Toronto, ON, Canada

Alicanto Mining Corp. has applied for two exploration licenses in the northwest of Bucaramanga, Colombia. The project encompasses disseminated gold mineralization in a marine basin. Within the area, the Client also owns an exploitation area contract of 7.65 km2.

This new gold district contains four sources of gold mineralization:

1. The active placers of the Lebrija River.

2. A paleo placer associated with a Jurassic marine basin (Witwatersrand-type) currently under exploitation

3. Gold associated with a sequence of Tertiary conglomerates

4. Gold associated with the bituminous-rich limestones (Carlin-type or MVT)

The other two potential sources of gold mineralization in the area are:

1. Gold associated with a potential Tonalite intrusive

2. A skarn type at the contact between the intrusive and the limestone.

The discovery resulted from a new model for the geological evolution of the region. It explains the richness of this new gold district. It also helped to design the best exploration method (GFcsa™[1]) to search for similar targets in the area.

The Company got all the permits and is currently exploiting the paleo placer. The current production is only 3.5 kg/month, but the company applied for an increment of 15 kg/month. We have defined 35.92 million cubic metres with a grade of 0.81 g/t which represents 58,916 kg of free gold (1,894,169 ozt). There are also 4.79 million cubic metres of magnetite with a grade of 1 g/t. For a 95% recovery, this represents an extra 4,548 kg of gold (146,229 ozt).

In summary, we have 63,464 kg of gold (2,040,398 ozt) of Inferred Mineral Resources in the exploitation license.

[1] Geoconsult Fenix complex system approach- survey includes geology, geochemistry, and geophysics simultaneously.

Hydrothermal Alteration and Gold Occurrence at the Paleoproterozoic Piaba Orogenic Gold Deposit, Maranhão State, Brazil

Deniro F. Gonçalves Costa[1], Gabriela V. Terenzi[1], Rosaline C. Figueiredo e Silva[1], Rafael G. Correa Lima[2], Gabe Kassos[3], Carlos Paranhos Jr[4]
1. Federal University of Minas Gerais (UFMG), Belo Horizonte, MG, Brazil, 2. MASA (Mineração Aurizona S.A), Godofredo Viana, MA, Brazil, 3. Equinox Gold Corp., Vancouver, BC, Canada, 4. Equinox Gold Corp., Belo Horizonte, MG, Brazil

The Piaba orogenic gold deposit, located in the São Luís cratonic fragment (Maranhão State, Brazil), is associated with the development of a subvertical strike-slip fault that cuts across Rhyacian metavolcano-sedimentary rocks (carbon-bearing schists, andesite/dacite, ultramafic rocks and felsic tuffs), quartz-diorite and granitoids (fine-grained, subvolcanic, granophyrice). The host metavolcano-sedimentary sequence formed in a subduction related setting, previously interpreted as the early-arc stage of an accretionary orogen (2240–2214 Ma) that was followed by voluminous subduction-related calc-alkaline arc magmatism (2168–2145 Ma) and by a collisional phase (2100 ± 15 Ma) that produced several peraluminous granite bodies.

This petrographic study was conducted on select diamond drill core samples, primarily quartz diorite ore host rocks from the Piaba deposit. Study goals include: (i) identify the mineral phases; (ii) describe the lithologic and mineralogical textures; (iii) estimate mineralogical composition; and (iv) characterize the associated hydrothermal alteration zone. Even though rocks are commonly intensively hydrothermally altered, primary features such as granophyric texture (in this study, an intergrowth of quartz and alkali feldspar) are present in most samples.

Hydrothermal alteration of the host rocks produced recognizable temporal and spatial zonations that include: 1. Early, distal to intermediate carbonate-chlorite-muscovite, commonly disseminated from veins-veinlets and with associated pyrite and which locally forms a stockwork texture and bleached zones; and 2. A proximal, ore-related, chlorite–sericite–carbonate–sulfide alteration. Pyrite is the dominant sulphide.

Gold occurs in: 1. Thin (millimetre to centimetre-scale), quartz-carbonate ± sulphide ± tourmaline bearing shear veins; and 2. Sub-horizontal, quartz-carbonate extensional veins. Native gold is rarely observed at wall rock-vein contacts. Increased vein density and sulphide abundance are the best indicators of gold mineralization.

The Carboniferous Shikebutai Iron Deposit in Western Tianshan, NW China: Petrology, Fe–O–C–Si Isotopes and Implications for Iron Pathways

Xiuqing Yang
Chang'an Univeristy, Xi'an, China

Submarine volcanic-hosted iron deposits are important sources of iron ore in NW China. Here we present the petrological characteristics and coupled Fe–O, C and Si isotope data on iron ores from the Shikebutai submarine volcanic-hosted hematite deposit in Western Tianshan region. Several stratiform and lenticular hematite-dominated orebodies occur in Carboniferous submarine volcano-sedimentary sequences in this region. The ores are mainly composed of hematite, quartz, and minor siderite with distinct alternating iron-rich and silica-rich bands. The hematite shows $\delta^{56}Fe$ and $\delta^{18}O$ values in the range of -0.31‰ to +0.80‰ and +2.2‰ to +7.0‰, respectively, and the jasper yields $\delta^{30}Si$ values of -1.90‰ to -1.20‰. Iron and Si were both derived from hydrothermal fluids related to submarine magmatism/volcanism. The Fe-bearing minerals in the Shikebutai deposit define distinct formation pathways. Hematite is the primary dehydrated Fe(III)-oxyhydroxide, and the Fe isotope data indicate fractionation resulting from the partial oxidation of Fe(II). The O isotope data reflect inheritance from the submarine hydrothermal fluids source. Jasper was produced during co-precipitation of silica adsorbed onto the Fe(III)-oxyhydroxides. The siderite-rich iron ore/volcaniclastic rock samples with a high and variable total organic carbon content (0.14%–5.57%), show negative $\delta^{13}C$ values (-3.0‰ to -1.1‰) and light $\delta^{56}Fe$ values (-1.11‰ to -0.84‰). Our isotope data, together with the common occurrence of hematite inclusions in siderite suggest that siderite was mainly produced by microbial dissimilatory iron reduction (DIR) during diagenesis. The geological, petrological, and isotopic data suggest that the Carboniferous Shikebutai deposit was precipitated through chemical and biogenic processes.

Fe Isotope Systematics Indicate Magmatic Origin of the Per Geijer Iron Oxide-apatite Deposits in Northern Sweden

Patrick Krolop[1], Yona Nebel-Jacobsen[2], Thomas Seifert[3]
1. Luossavaara Kiirunavaara AB, Kiruna, Sweden, 2. Monash University, Melbourne, VIC, Australia, 3. TU Bergakademie Freiberg, Freiberg, Germany

Iron oxide-apatite (IOA) deposits of Kiruna-type are key sources of iron ore in Europe. They are sourced from the northern regions of Sweden (Kiruna – Svappavaara – Malmberget). The Per Geijer IOA deposits near Kiruna display significant exploration targets for the Swedish mining company LKAB and are currently being reassessed for possible future mining. Although studied to a lesser degree compared to the well-known Kiirunavaara deposit, they can provide new insights into the controversially discussed origin of IOA deposits. Ore-forming scenarios span from low-temperature hydrothermal processes to high-temperature origin from magma or magmatic fluids. A set of new iron isotope data of magnetite in Kiruna-type iron ores from Per Geijer and Kiirunavaara, and comparative data from two skarn-type deposits of northern Sweden, Kaunisvaara and Masugnsbyn were obtained and compared with available global reference data from layered intrusions, volcanic provinces, and low-temperature and hydrothermal iron ores. The massive magnetite from the Per Geijer deposits exhibit $\delta^{56}Fe$ values of 0.18 to 0.44‰ that overlap with magnetite data from layered intrusions and recent volcanic provinces (≥ 800 °C) and corroborates with previous data from IOA deposits in the Kiruna district. In contrast, sampled magnetite from skarn-type deposits in the region (-0.21 to 0.09‰) correspond dominantly with published reference suites of hydrothermal magnetite formed by hydrothermal fluids at temperatures ≤ 400 °C. Therefore, a magmatic derivation is proposed for magnetite of the Per Geijer deposits. Despite some available literature data, this needs to be further evaluated by corresponding O-isotope information. Analyzed $\delta^{56}Fe$ data of hematite from the Per Geijer deposits yield similar values of 0.24 to 0.32‰. Since most of the hematite present in these deposits results from direct replacement of magnetite it is suggested that the Fe isotopic ratio is not significantly affected by oxidation events.

Geology and Geochemistry of the Namicupo Gold Prospect, Mozambique Belt, Northeastern Mozambique: Insights from Ore Mineralogy, Fluid inclusions and Stable Isotopes

Manuel Nopeia[1], Akira Imai[1], Ryohei Takahashi[2], Kotaro Yonezu[1], Daúd Jamal[3], Andrea Agangi[2], Pearlyn Manalo[2], Thomas Tindell[1], Hinako Sato[2]
1. Department of Earth Resources Engineering, Kyushu University, Fukuoka, Japan, 2. Department of Earth Resource Science, Akita University, Akita, Japan, 3. Department of Geology, Eduardo Mondlane University, Maputo, Mozambique

The Namicupo gold prospect, hosted by the Mozambique Belt, northeastern Mozambique, occurs as quartz veins parallel to the foliation of metasedimentary rocks. This study discusses the ore mineralogy, fluid inclusion, and stable isotope characteristics of the unreported Namicupo gold mineralization. Based on petrographic observations, the ore in the Namicupo prospect is divided into two stages. The first stage is represented by native gold associated with pyrite and chalcopyrite. The second stage is interpreted as supergene mineralization, characterized by electrum intergrown with barite and oxyhydroxides, which replaced primary sulfides. Primary quartz from the mineralized veins hosts three types of fluid inclusions that coexist in the same crystal. Type A inclusions consist of three phases (aqueous liquid, CO_2 liquid, and vapor), with melting and homogenization temperatures of CO_2 varying from -54.8 to -51.2 °C and +26.7 to +30.5 °C, respectively, and salinity of 1.0 – 7.8 wt.% NaCl eq. Type B inclusions are single-phase CO_2 liquid at room temperature, with melting temperature between -54.8 and -53.9 °C; the homogenization temperature of inclusions exhibiting vapor nucleation during cooling ranges from +12.9 to +29.3 °C. Type C two-phase (liquid and vapor) aqueous inclusions show variable degrees of vapor filling, with homogenization temperature between 209 and 337 °C, and salinity of 4.0 – 10.8 wt.% NaCl eq. The oxygen isotopic ratio ($\delta^{18}O_{SMOW}$) of water calculated from that of primary quartz, and sulfur isotopic ratio ($\delta^{34}S_{CDT}$) of primary sulfides vary from -2.6 to +0.9 ‰ and -3.5 to +0.9 ‰, respectively. The primary mineralization at the Namicupo prospect is classified as an orogenic-type gold deposit affected by supergene alteration under a moderately oxidizing environment. Primary gold mineralization was caused by an aqueous-carbonic metamorphic fluid at approximate temperature and pressure conditions, estimated from intersecting isochores of aqueous and carbonic inclusions, of 320 °C and 170 MPa.

Karst-Hosted MVT Pb–Zn Deposits in Fold-thrust Systems: a Case Study of the Changdong Deposit in the Sanjiang Belt, China

Yingchao Liu, Longlong Yue
Institute of Geology, Chinese Academy of Geological Sciences, Xicheng District, China

Karst is the most common ore-controlling structure in Mississippi Valley-type (MVT) Pb–Zn deposits. Unfortunately, little is known about the formation process of karst caves containing MVT ores and their associated Pb–Zn mineralization in fold-thrust belts. The Changdong MVT deposit is hosted by karst caves located in the fold-thrust system of the Sanjiang metallogenic belt, Tibetan Plateau. It's a natural laboratory for studying the question above. The deposit is situated in the Late Permian limestones and hosted by a large-scale breccia belt in the hanging wall of a regional thrust fault. The breccia belt is composed of breccias and internal sediments. Breccias consist of limestone fragments within a clastic matrix. Internal sediments, consisting of K–Al silicate clays, quartz, and rock and calcite fragments, fill the voids between the breccias. According to geological observation, C-O isotope analysis, detrital zircon U-Pb dating, and palynological geochronology work, these breccias and internal sediments were the major karst factors in a meteoric paleo-karst system forming during the regional compassion. Pb-Zn sulfides precipitated in the deposit by replacing the carbonates or filling the voids in the internal sediments. They have $\delta^{34} S_{V\text{-}CDT}$ values from -16.9 ‰ to 15.8 ‰, indicating a preexisting H_2S reservoir resulting from bacterial sulfate reduction (BSR). A three-stage process for the Changdong deposit was proposed. Firstly, Permian limestone was uplifted to the near-surface by the regional thrusting during the India-Eurasia collision, and karst caves were formed by meteoric dissolution. Secondly, a reduced sulfur trap formed in the paleokarst caves by BSR during regional compression. Thirdly, metal-rich fluids migrated into the Changdong deposit through tensile faults formed during the transition from compressive stress to extensional strike-slip stress, and the sulfides precipitated. This study provides new data to establish a consistently geological framework for evaluating karst caves and associated MVT Pb–Zn mineralization in fold-thrust systems.

A Study of Mineralogy of Pyrite and Au Mineralization in Micro-Disseminated Gold Deposit in Southwest Guizhou, China

Huan Jiang[1], Aihua Xi[1], Baojin Zhao[1, 2]

1. Southwest Petroleum University, Chengdu, China, 2. Department of Environmental Sciences, University of South Africa, Johannesburg, South Africa

The micro-disseminated gold deposit in Southwest Guizhou is a typical medium-low temperature hydrothermal deposit while the strata-bounded nature of Au mineralization hasn't been clearly understood. We analyze the morphology, composition and distribution of pyrite and the occurrence of Au of the Shuiyindong gold deposit through mineralogy, petrography, EPMA, ESEM and other high-precision in-situ test methods. The results indicate that there are at least three types of pyrite: coarse, fine and framboidal pyrite. The coarse pyrite has a regular core and 2-3 outer part of the secondary enlarged ring belt, which the elements such as S, As and Au show an annular distribution, indicating that the inner and outer zones were formed in different hydrothermal metallogenic stages. The fine pyrite is mostly symbiotic with arsenopyrite and has high content of Au and As. The framboidal pyrite is filled in the biological skeleton (Ty1) and microfracture (Ty2) as aggregates, which is closely associated with organic matter and almost contains no Au and As.

Gold occurs as invisible ions or atoms in lattice or as native micron particles. The former mainly exists in As-rich fine pyrite and the enlarged ring of coarse pyrite while the latter are distributed in vugs, microcracks of the host rock, pyrite, or organic matter, showing typical fracture-controling characteristics.

The above research on the pyrite shows that this type of gold deposit has the characteristics of multiple overprinting of hydrothermal activities. Furthermore, the structures may provide a conduit for hydrothermal migration while the coexisting of abundant bioclasts in the orebody may play an important role in the strata-bound characteristics of micro-disseminated gold deposit.

Preliminary Analysis of Zn and Ni Sulfides in Calcareous Black Shales, Puyango, Ecuador

John L. Manrique[1, 2], José E. Ortiz[2], Rastislav Milovsky[3], Tomáš Mikuš[3], Sergii Kurylo[3]
1. Geosciences Department, Universidad Técnica Particular de Loja, Loja, Ecuador, 2. Universidad Politécnica de Madrid, ETSI Minas y Energía, Madrid, Spain, 3. Earth Science Institute of the Slovak Academy of Sciences, Banská Bystrica, Slovakia

The mineral occurrence of Puyango, Ecuador, contains high values of several critical metals: vanadium (mean 2051 mg/Kg; maximum 7220 mg/Kg), zinc (mean 939 mg/Kg; maximum 5610 mg/Kg), nickel (mean 255 mg/Kg; maximum 1160 mg/Kg), among others. It is hosted in black calcareous shales and black limestones from the Puyango Formation of Cretaceous age. Analysis of selected samples were carried out by Electron Probe Microanalysis (EPMA), identifying Fe, Zn, Ni and Cu sulfides: pyrite (including framboid pyrite), sphalerite, millerite and chalcopyrite (in smaller quantities). Preliminary results indicate anoxic to euxinic conditions during sedimentation, according to elemental ratios in bulk composition of black calcareous shales and black limestones by ICP-MS: ($U/Th > 4.8$), ($V/(V+Ni) > 0.6$), ($V/Cr > 4.8$).

Characterization of the Gold Deposits Associated with the Aurumina Granitic Suite and Ticunzal Formation in Northeastern Goias State, Central Brazil

Gabriel A. Barbosa Silva, Nilson Francisquini Botelho
University of Brasilia, Brasilia, DF, Brazil

The studied region is situated in the northeast portion of the Goias State, central Brazil, where there are several gold occurrences sporadically exploited since the XVIII century. The region is located in the northern part of the Brasilia fold belt, in its exposed Paleoproterozoic basement rocks. The gold mineralizations are related to Paleoproterozoic peraluminous, syn-tectonic granites of the Aurumina Suite, and the metasedimentary, graphite-bearing country rocks of the Ticunzal Formation. The goal of this work is to characterize the mineralizations and their controls and discuss the roles of the granitic and country rocks as sources of metals and fluids. The ore is hosted in quartz veins associated with hydrothermally altered muscovite-quartz mylonites and phyllonites in shear zones near the contact between the 2.15 Ga biotite-muscovite granite of the Aurumina Suite and the graphite-bearing paragneiss and schist of the Ticunzal Formation. The mineralization is synchronous with the granitic intrusion, with syn-emplacement shearing and hydrothermal alteration. Hydrothermal alteration resulted in intense silicification close to the contact between the veins and the host rock, as well as the crystallization of hydrothermal muscovite, chlorite, and carbonate. Primary sulfides in the ore include galena, pyrite, chalcopyrite, arsenopyrite, and pyrrhotite, along with secondary copper minerals. Gold occurs along fractures on quartz, as fine inclusions on arsenopyrite and disseminated in the quartz veins. Data from arsenopyrite geothermometer and sulfur isotopes indicate temperatures in the range of 200°C to 300°C for the gold mineralization. Despite having similarities with orogenic gold deposits, the mineralization has some essential features more consistent with the intrusion-related gold deposit model. Such features include its cordilleran hinterland tectonic environment, a regional association with tin and tungsten mineralizations, and a strong association between the ore and the granitic pluton, with structures active during emplacement and cooling.

A Potential Cu Skarn in the Huachocolpa District-Peru

Ivan E. Branes, Miguel Miranda, Mario Rosas, Jim Torres, Omar Yaringano
KOLPA MINING COMPANY, LIMA, Peru

The escondida project it is located at 7 km to the South West in the district of Huachocolpa, Huancavelica, Peru. it belongs to the Kolpa mining company and is a potential target of skarn.

The local geology has an outcrop of the Casapalca formation of the upper cretaceous, tantara formation, Apacheta formation, two intrusions dioritic and dacitic composition. The Casapalca formation is composed of 4 lithostratigraphic units recognized in the area. unit 3 shows calcareous levels and red shale intercalations on the surface there are recrystallized calcite veins product from hydrothermal processes.

Geophysical works such as IP, magnetometry and resistivity were carried out, obtaining anomalies in each one of them, adding the lithological variable of the calcareous horizon of the Casapalca Formation and the 3 mains, were solid arguments to carry out a drilling campaign with the objective of finding a Skarn deposit.

In 2021, a first drilling campaign of 1000 meters was programmed, recognizing 3 main veins.

There are three main veins: Camila, Camila I and Sr Manue with strike NW-SE and dip to the SW. In surface show minerals such as sphalerite, galena and freibergite.

In the drillhole ESC-21-04, at 180 meters below surface it intercept Camila vein as clast soported breccia with mineralization of fraibergite, sphalerite and chalcopirite; calcite, barite and brown Garnet in the matrix, with the following results: @2.60 meters, 0.07 Au g/TM, 7.98 AgOz/TM, 0.39% Cu, 0.72 Pb, 0.23% Zn.

A second drill hole program is currently underway to a depth of 800 meters with the objective of finding and defining a skarn deposit with mineralized zones greater than 1M MT at 3% Cu equivalent.

Deciphering Conglomerate-host Gold Mineralisation in Moto Greenstone Belt, Congo Craton, DRC: Implication for Gold Exploration

Yann Mpaka Waku[1, 2], Bjorn von der Heyden[1], Trézène Bampata[2], Fillsmith Luzolo Ndongani[3, 2], Etienne Mwandale[2]
1. Stellenbosch University, Stellenbosch, South Africa, 2. Barrick Gold Corporation, Toronto, ON, Canada, 3. University of Cape Town, Cape Town, South Africa

The giant Karagba-Chauffeur-Durba (KCD) gold deposit and its associated peripheral gold deposits are located within the Neoarchean Moto Greenstone Belt in the Kibali district in the northeastern portion of the Congo Craton, Democratic Republic of Congo. These gold deposits have been defined as orogenic-type gold deposits with mineralisation mainly controlled by structure, through which hydrothermal fluids have transported and precipitated gold, notably in the local Banded Iron Formations (BIF) where chemical trapping mechanisms are expected to control gold siting and tenors. Although volumetrically-important for their role as host lithologies, less altered clastic sediments (mainly conglomerates), have received less scientific and exploration attention, despite displaying high gold grades in places (i.e., >100 $g.ton^{-1}$). To fill this knowledge gap, we present a comprehensive characterisation of conglomerate-host gold in the Kibali gold district, considering samples from KCD, Sessenge, Gorumbwa, Kombokolo, Agbarabo, and Rhino deposits. This investigation considered an extensive data set comprising polished samples from several boreholes, over nine million gold assay values distinguished according to host lithology, microscopic studies of 3824 individual gold grains, and mineral chemistry using Scanning Electron Microscope analyses on 397 gold grains. The identified gold grains show a variety of morphological forms with some notable variations in their mineral chemistry. Textural associations range between attachment to sulphide mineral surfaces, inclusions within sulphides, and as discrete grains showing only random association with the silicate gangue. Over 53% of all mineralised core intersections comprise conglomeratic lithologies, emphasising this unit's importance as an essential gold host and highlighting the importance of renewed scientific interest in the reasons for its endowment. These reasons will be discussed in the context of regional geology, the host rock's physico-chemical properties, and the implications for exploration paradigms within the broader KCD area.

Swir Analysis of Alteration Assemblages Associated with Porphyry-hosted Gold Mineralization of the Yixingzhai Deposit in the Taihangshan District, North China Craton

Wen-Sheng Gao[1, 2], Jiacheng Liu[3], Xiao-Dong Deng[1], Li-Zhong Zhang[4], Jian-Wei Li[1, 2]
1. State Key Laboratory of Geological Processes and Mineral Resources, China University of Geosciences, Wuhan, China, 2. School of Earth Resources, China University of Geosciences, Wuhan, China, 3. Department of Earth Sciences, The University of Hong Kong, Hong Kong, China, 4. Zijin Mining Group Co Mineral Exploration Institute, Xiamen, China

Shortwave infrared (SWIR) reflectance spectroscopy has been widely used to characterize phyllosilicate alteration assemblages associated with ore deposition, and thus may shed significant light on ore genesis and provide guides for mineral exploration. Porphyry-hosted gold mineralization has been recently discovered in the Yixingzhai deposit of the Taihangshan district, North China Craton, where quartz-sulfide lodes have been mined for three decades. It consists of pyrite disseminations and minor thin veinlets in the Early Cretaceous Hewan granite porphyry cut by an NW-trend fault (locally termed F4), with proven reserves of 54.7 t gold at an average of 2.12 g/t. Gold mineralization is associated with extensive hydrothermal alteration. In this study, more than 900 samples from 29 diamond drill holes along three cross-sections that penetrate main orebodies were collected and measured using an Analytical Spectral Devices spectrometer to characterize the alteration assemblages. The results indicate that the ore-related hydrothermal alteration mainly consists of illite, phengite, and muscovite. The wavelength of the Al-OH band (2200 wvl) of those minerals ranges from 2199 nm to 2217 nm, but mostly between 2204 nm and 2212 nm. The 2200 wvl values for these phyllosilicates are negatively correlated with the contents of octahedron Al in the minerals which is a parameter of the pH of the ore-forming fluids. Furthermore, values of 2200 wvl increase outwards from the F4 fault, indicating its role as the structural conduit for the ore fluids. The illite spectral maturity (ISM) for the phyllosilicate minerals ranges from 0.4 to 9.2, with high-ISM samples (ISM > 1) containing gold higher than 0.1 g/t. Results presented here highlight the successful utilization of SWIR reflectance spectroscopy in mineralization and exploration of disseminated gold deposits hosted by igneous rocks.

New Constraints on the Formation of the Bolcana Au-Cu Porphyry System (Apuseni Mountains, Romania)

Sava Markovic[1], Jean-Marc Benz[1], Tim Baker[2], Maria P. Castellanos Melendez[1], Paul Ivascanu[3], Thomas Driesner[1], Albrecht von Quadt[1]
1. ETH Zurich, Zurich, Switzerland, 2. Eldorado Gold Corporation, Vancouver, BC, Canada, 3. Dundee Precious Metals, Sofia, Bulgaria

The inferred resource of the Bolcana Au-Cu porphyry deposit (Apuseni Mountains, Romania) has recently been estimated to 381 Mt at 0.53 g t^{-1} Au and 0.18% Cu, containing 6.5 Moz of Au and 686,000 t of Cu. Here, we build on extensive drillcore logging and petrography to describe the Bolcana porphyry system along a continuous profile down to ~2 km depth. We track the magmatic evolution using zircon U-Pb and trace-element data, and show that the bulk of Au-Cu mineralization in Bolcana occurred at low temperatures (<410 °C) in the waning stages of the magmatic-hydrothermal system.

We group the calc-alkaline magmatism in Bolcana into three stages: dacite porphyry (stage I) first intruded at ~13 Ma, and ore-related, andesitic stage–II and stage–III porphyries followed at ~11 Ma. Despite their overlapping LA-ICP-MS and TIMS zircon dates, field evidences unambiguously demonstrate that stage–III porphyries intruded the previous lithologies in the form of narrow dikes, causing widespread brecciation while the hydrothermal system was still active. Trace-element composition of zircons from ore-related porphyries suggest their crystallization in the presence of apatite and hornblende, over a temperature span between ~675–950 °C.

Veining comprises quartz veins of A, AB, B, and banded ("Maricunga") style, C and D veins, and three types of magnetite veins. Sulphide ore minerals are, present in interstices in quartz veins and hostrocks as well as in sulphide-only C veins, and all texturally post-date quartz precipitation. Ti-in-quartz and fluid inclusion microthermometry constrain quartz precipitation to a continuous temperature range from 410–530 °C and 500–700 °C, respectively, implying that the sulphides were precipitated at even lower temperatures. Vein quartz in all stages precipitated from a purely magmatic fluid ($\delta^{18}O$=7.6–8.9‰). The predominance of low-density fluid inclusions within different quartz vein types suggests that quartz vein formation predating sulphide precipitation likely occurred under low pressures.

Genesis and Geochemistry of the Pahardiha- Rungikocha Gold Deposit in North Singhbhum Mobile Belt, Eastern India

Anmol Barla[1], Sahendra Singh[1], Rajarshi Chakravarti[2], Ramesh C. Behera[1]
1. INDIAN INSTITUTE OF TECHNOLOGY (INDIAN SCHOOL OF MINES), DHANBAD, INDIA, DHANBAD, JH, India, 2. INDIAN INSTITUTE OF TECHNOLOGY, ROORKEE, ROORKEE, UK, India

The Pahardiha- Rungikocha gold deposit is situated in the Chaibasa Formation of the Singhbhum Group within the Proterozoic North Singhbhum Mobile Belt (NSMB), in eastern India. The total gold ore reserve reported from the deposit is 0.487 million tonnes, with an average grade of 3.43 ppm. Field observations, coupled with petrographic studies, shows three-phase of deformations, i.e., D1, D2, D3. The deformation D1 and D2 represent the progressive regional metamorphism, and D3 shows the displacement fractures. The post-peak-metamorphic, i.e. D3 deformation, is associated with the gold mineralization within the quartz-chlorite schist. In this deposit, gold occurs as native specks hosted within the fractures network and along the grain boundaries of euhedral pyrite. The gold-bearing pyrite is coeval with the chromite cored magnetite (Mag-II), where euhedral magnetite rims show epitaxial overgrowth over the relict chromite cores. The Major oxide geochemical classification diagram indicates that the host rocks source belongs to wacke to Fe-Shale. The host rock geochemistry suggests the mixed source of felsic to mafic rocks. The critical element ratios indicate that the sediments have undergone moderate to high degrees of chemical weathering. The REE data reveal the enriched LREE while HREE shows flat to slightly depleted pattern. The REE shows the Eu negative anomaly, which indicates the feldspar fractionations. The chemistry of chromite cored magnetite (Mag-II) and host rock chemistry reveals that the host rock holds the primary control for the chemistry of chromite cored magnetite (Mag-II).

The Geological Setting and Hydrothermal Alteration of the Tucano Gold Deposit, Guiana Shield, Brazil

Gabriel A. Soares[1], Rosaline C. Silva[1], Steffen G. Hagemann[2], Lydia M. Lobato[1, 3], Rogerio A. Lucena[4]
1. Graduate Program in Geology, Institute of Geosciences (IGC), Federal University of Minas Gerais, Belo Horizonte, MG, Brazil, 2. Centre for Exploration Targeting, University of Western Australia, Perth, WA, Australia, 3. Hydro Fluids & Minerals, Rio de Janeiro, RJ, Brazil, 4. Great Panther Mining Limited, Pedra Branca do Amapari, AP, Brazil

The Tucano gold deposit in the southeastern Guiana Shield, northern Brazil is currently operated by Great Panther Mining. Total mineral reserves are estimated to be 9.7 Moz at 2.00 g/t Au (www.greatpanther.com/operations/producing-mines/tucano-mine/). The deposit is hosted in regionally metamorphosed Paleoproterozoic supracrustal rocks, dominantly of metasedimentary origin. Stratigraphic units include banded iron formations (BIF), metacarbonate, calc-silicate and metapelitic rocks. Gold-related hydrothermal alteration is structurally controlled by a N-S striking, steeply dipping brittle-ductile shear zone controlling orebody geometry and location. Alteration mineralogy varies with host rocks, but in most metasedimentary rocks serpentine, talc, chlorite, mica, carbonate and minor amphibole define a distal alteration zone. The proximal zone is characterized by mica, carbonate, amphibole, clinopyroxene and locally garnet. Hydrothermally altered BIF lack clinopyroxene, whereas metacarbonate rocks almost ubiquitously contain coarse grained clinopyroxene. Green mica is characteristic in altered BIF and vector high-grade (> 3 g/t) gold zones. Hydrothermally altered rocks may contain either one or both of the following amphiboles: a Ca-poor (cummingtonite-grunerite) and a Ca-bearing (tremolite-Fe-actinolite) species. The former is closely associated with silicification processes, typically involving Fe addition, whereas the latter is typical in the hydrothermal alteration of the metacarbonate rocks. Overall, hydrothermal alteration zones are characterized by sharp transition from least altered to intensively altered rocks in the ore zones. Locally, granite dikes display thin (< 5 cm) garnet-rich hydrothermal alteration haloes. Sulfides in all host rocks comprise mainly pyrrhotite, which is locally overprinted by pyrite, and rare chalcopyrite and arsenopyrite. In BIF, magnetite bands are replaced by pyrrhotite, which may also be observed as irregular masses. The replacement of magnetite by pyrrhotite typically marks the transition from distal to proximal hydrothermal alteration zone in BIF. The hydrothermal alteration zoning in the Tucano deposit is compatible with other observed high-temperature orogenic gold systems in Brazil and in granite-greenstone belts worldwide.

Ore Characteristics and Mineralogy of Fe-Sn Skarn Deposit in Batubesi Area, East Belitung, Western Indonesia

Wahyu V. Pratama, Andika Artyanto, Nur R. Nabawi, Anton Murtono
PT Timah Tbk, Pangkalpinang, Indonesia

Batubesi skarn deposit is one of the largest primary tin resources in Indonesian Tin Island. Several methods have been carried out to identify ore characteristic and mineralogy of tin-bearing skarns. The result showed that garnet and clinopyroxene was the Batubesi skarn type mineral. In addition, minor scheelite was also found to be the tungsten-bearing minerals. Cassiterite was the main valuable ore mineral and it was found as finely disseminated and predominately observed locked to iron oxide minerals, particularly magnetite and hematite. The extremely fine disseminated cassiterite in this deposit cannot be liberated for physical beneficiation techniques. Sn concentrations between 0.2% and 0.5% can be recognized from the abundance of clinopyroxene by the appearance of minor garnets, both banded and non-banded skarns. Moreover, oxidation process on the surface which is characterized by the presence of iron oxide minerals causes higher Sn grade due to enrichment by leaching process. Thus, the Sn and Fe concentrations in clinopyroxene can be used to evaluate skarn deposit's potential, which is helpful in exploration at Belitung Area.

Characterization of Multiepisodic Events of Calcatreu Project, Rio Negro Province, Argentina

Matías S. Wernert
Instituto de Recursos Minerales - Universidad de La Plata, Ciudad de La Plata, Argentina

The Calcatreu project is a low sulphidation Au-Ag deposit located in Rio Negro province, Argentina. Exploration operations are carried under the management of Patagonia Gold SA since its acquirement in 2018. Indicated resources are estimated at 746 kOz, with an ore grade of 2.36 g/t AuEq.

The deposit encompasses multiple arrays of NE-SW trending quartz and carbonate veins, hosted by jurassic andesites, dacites and tuffs of Taquetrén Formation. These rocks are the product of the many extensional events that occurred during the Godnwanide Orogeny. Furthermore, the mineralization is estimated to have an upper Jurassic age, based on comparisons with other similar deposits of the Patagonia.

A paragenetic and mineralogical study was done in the main veins of the deposit: Veta 49, Nelson Este, Nelson Oeste, Castro and Castro Sur. These veins are composed mostly by multiepisodic mineralized events of quartz and carbonate with banded and breccia textures. Many quartz variants have been identified, with white, grey and colourless being the most common, but other types such as pink, black, green and amethyst have also been found. Carbonate minerals are mostly white crustiform and banded colloform calcite, but additional pink, grey and brownish varieties have also been recognized.

The veins have a small amount of sulphide minerals, to which gold is commonly associated. Pyrite, chalcopyrite, galena and red-brown sphalerite have been identified as occasional thin veinlets or as clasts within hydrothermal breccias. Ginguro layers are somewhat common alongside other silica bands, usually showing signs of deformation, such as sigmoidal banding and shingle-type brecciation. Sulphosalt species may be present also, given the elevated As and Sb concentrations commonly present in sulphide-rich drill samples.

Calcatreu deposit represents a major project in the Somún Curá Massif, a sub-explored region that presents similarities with the important low sulphidation Deseado Massif province.

Geological Characterization of Auriferous Sulfide Within the Vempalle Dolostone, Cuddapah Supergroup, Southern India: Implication for Carlin Like Gold Occurrence

Ramesh C. Behera, Sahendra Singh, Anmol Barla
INDIAN INSTITUTE OF TECHNOLOGY(INDIAN SCHOOL OF MINES) DHANBAD, DHANBAD, JH, India

The Earth's mineral resources are limited in a world of ever-increasing demand from commerce and technology. As a result, new frontiers in mining metals and energy resources are necessary for future economic and technological advancement. Nanotechnology and relevant industrialization appear to be high on humanity's development agenda.

The study area for the present work belongs to the Vempalle Formation within the Cuddapah Supergroup in southern India. The dolostone within the Vempalle Formation is well known for its uranium mineralization. However, our preliminary work also shows the presence of auriferous sulfides within the uraniferous dolostone. Gold within the formation is associated with pyrites and arsenopyrites. Petrographic study reveals several types of pyrite, i.e., diagenetic, hydrothermal, and framboidal aggregate types. The gold values within these sulfides range from 10 to 1500 ppm. Arsenic zoning within the pyrite crystal indicated the transportation from earlier digenetic pyrite and precipitated onto the later euhedral hydrothermal ones. Major oxide co-relation indicates the typical detrital origin. Major and trace elements systematics suggest the anoxic depositional environment of the mineralized dolostone. The Pb value indicates the biogenic activities which provided the anoxic environment for sulfide enrichment. The negative Eu anomaly from the REE plot means the host dolostone is recycled from the upper continental crust. The Gd/Yb values confirm the post-Archean depositional environment. The Au-rich domains exhibit Au (Sb), (Tl), (Hg), and (Cu) peaks in the energy dispersive X-ray (EDX) spectrum. A positive correlation between Au and As indicates that they are co-precipitated. These observations confirm the presence of invisible gold both as nano inclusions and solid solution within the crystal structure of arsenian pyrite and suggest the Carlin-type of gold occurrence within the Vempalle dolostone of the Cuddapah Supergroup in southern India.

VMS Targeting Challenges: Rethinking the World-class IPB Province

Filipa Luz[1], João Gonçalves[1], Joana Fragoso[1], Higino Tavares[1], Nídia Sousa[1], Scott Napier[2], Stanislawa Hickey[2]
1. ALMINA, Aljustrel, Portugal, 2. Mira Geoscience, Quebec, QC, Canada

The Iberian Pyrite Belt (IPB) is a world-class VMS metallogenic district. The energy transition and the demand for critical raw elements such as Cu-Zn have led to renewed exploration campaigns in the IPB. The historical track record shows that improvements and changes in conceptual models, along with advances in geophysical methods, 3D geological modelling and a detailed geochemistry approach were critical for recent discoveries such as La Magdalena (2013, Spain) and Sesmarias (Portugal, 2014). The Albernoa exploration license is a greenfield project on the IPB that is located near Aljustrel (Cu-Zn active mine in Portugal), which has been the focus of some studies by several companies during the last years. Within the Albernoa exploration license, new methodologies recently developed for IPB have been tested to drive the exploration campaign. The combination of high resolution geological, mineralogical, geochemical and geophysical data allowed the identification of some anomalies in depth that are currently being tested by drill-holes. Airborne ZTEM and ground-based fixed loop SQUID TEM data identified EM responses with similar amplitudes to known massive sulfide accumulations in active mines, and which were not yet tested by drilling. Compilation of new and historical geophysical data and reprocessing and modelling using state-of-the-art techniques, such as 3D inversion of electromagnetic data, will yield better and more focused geophysical targets than previously available. For the Albernoa area, specific Alteration (AI) and Mineralization Indexes were developed based on micro-analytical studies. Metapelites record values of (Cu+Zn+Pb)/Sc and (As+Sb)/Sc > 10, which allow to separate barren from altered/mineralized sequences. The same is observed for metavolcanics where AI_{major} (Fe_2O_3+MgO+MnO/Fe_2O_3+MgO+MnO+K_2O+Na_2O) varies between 0.25 to 1 and AI_{trace} (Zn+Cu+As+Sn/Zn+Cu+As+Sn+Rb+Ba) > 0.4, suggesting hydrothermal alteration and mineralization. VMS targeting is a challenge and for new discoveries it is crucial to gather and model from macro to nanoscale.

Nature and Paragenesis of Copper Mineralization of the Viscaria Property, Kiruna District, Northern Sweden

Koray Tasbicen[1], William X. Chávez, Jr.[1], Marcello Imaña[2]
1. New Mexico School of Mines, Socorro, NM, USA, 2. Copperstone Resources, Kiruna, Sweden

The Norrbotten Region in Northern Sweden is recognized by its mining history for Fe-, Cu- and Au-, hosted within Paleoproterozoic-age rocks. Located 5km from the giant Kiruna IOA deposits, the Viscaria deposit was mined for copper in the 1980s. Copper mineralization at Viscaria comprises mantos distributed along several volcano-sedimentary horizons, that host massive to semi-massive magnetite and carbonate with generally disseminated pyrite – pyrrhotite – chalcopyrite – sphalerite.

Evolution of the deposit starts with pervasive and extensive biotite alteration followed by development of magnetite, calcic amphibole, chlorite and talc alteration. Two generations of paragenetically early, disseminated pyrite are replaced by pyrrhotite. Copper sulphide mineralization consists of semi-massive, disseminated and veined chalcopyrite; chalcopyrite is locally replaced by high-Fe sphalerite. Latest alteration comprises carbonate pyrite veins with vuggy textures and pervasive carbonate development.

To help understand the distribution of trace elements within Viscaria, ore and ore-related minerals including amphiboles, have been analyzed for specific trace elements. Preliminary results show that the later generation of pyrite is enriched in Co; pyrrhotite is enriched in Co and Ni. Sphalerite shows slight enrichments in Cd and, locally, in Co. Characterization of trace elements in the ore and alteration minerals along with textural and paragenetic observations will help to understand the evolution of the Viscaria hydrothermal system and its significance for exploration within the Kiruna District.

The Importance of Bi-Te Nanoparticles in the Remobilization of Polymetallic (Au-Ag-Te-Bi) Orogenic Gold Mineralization

Michael Herzog[1], Crystal LaFlamme[1, 2], Laura Petrella[2], Bertrand Rottier[1], Georges Beaudoin[1]
1. Université Laval, Québec City, QC, Canada, 2. The University of Western Australia, Perth, WA, Australia

Recent petrographic and microthermometric observations in orogenic gold systems imply that low-melting-point chalcophile elements can co-exist in the form of liquid-polymetallic Au-Ag-Te-Bi-melt-droplets with hydrothermal fluids to temperatures down to 135°C. These melts can efficiently scavenge Au from Au-undersaturated fluids through adsorption-reduction mechanisms along the polymetallic droplet-surface-area and, on further cooling, form Au-Ag-Te-Bi-minerals, visible as solid inclusions in pyrite, observed in various orogenic gold districts. Orogenic gold deposits commonly experience multiple hydrothermal events, which have the potential to remobilize polymetallic inclusions. However, their remobilization potential, mode of transport and chemical evolution remain elusive. Orogenic gold mineralization in the Abitibi (Canada) is distributed along the E-W striking Larder-Lake-Cadillac-fault-zone and hosted in lower-order structures that cut volcanic rocks and intrusions of the greenstone belt. These lower-order structures comprise brittle-ductile reverse shear zones and sub-horizontal tension veins, composed of quartz-tourmaline-carbonate. The shear-zones record two hydrothermal events 1) at ~2643 Ma, introducing the majority of gold as native Au and Au-Ag-Te-Bi-inclusions within pyrite; 2) a younger event at ~2607 Ma replaces sulfides and precipitates carbonate-chalcopyrite-native Au and carbonate-Au-Ag-Te-Bi-phases in pyrite fractures.

Here, we investigate the mineral-scale gold distribution of the Goldex deposit (Abitibi) by examining the potential of Au-Ag-Te-Bi-remobilization from pyrite-hosted, µm- to nm-sized polymetallic inclusions into texturally late sulfide fractures. Preliminary transmission electron microscopy (TEM) and nano-scale energy dispersive spectroscopy of two foils, extracted from texturally late pyrite fractures, hosting calcite-Au-Ag-Te-Bi-assemblages, reveal that infill of amorphous carbon forms layers, <300 µm-thick, at the interface between Te-Bi-phases, adjacent Au, calcite and pyrite host. The amorphous carbon phase contains sub-angular to rounded inclusions of Te-Bi nanoparticles between 10 to 50 nm. This suggests that late, reduced hydrothermal fluids dissolved pyrite-hosted Au-Ag-Te-Bi-inclusions and were remobilized along pyrite fractures. Our data imply that naturally occurring Te-Bi nanoparticles play a crucial role during the hydrothermal remobilization of already-present Au-Ag-Te-Bi-inclusions.

Element Fluxes During Alteration in High-sulfidation Epithermal Systems

Ethan R. Tonks[1, 2], Jamie J. Wilkinson[1, 2], Robin N. Armstrong[1], Andrew T. Wurst[3]
1. Department of Earth Sciences, Natural History Museum, London, United Kingdom, 2. Department of Earth Science and Engineering, Imperial College, London, United Kingdom, 3. Barrick Gold Corporation, Toronto, ON, Canada

The processes involved in the alteration of rocks associated with high-sulfidation epithermal ore deposits, which host significant quantities of gold and silver, cause a large-scale and complex migration of chemical elements. Elements can either be added into the system by magmatic-hydrothermal fluids or mobilized from one alteration facies to another due to the hydrothermal dissolution of minerals in the precursor rock and subsequent re-precipitation as new minerals elsewhere.

In this study, we present an elemental mass transfer analysis of alteration associated with the Pueblo Viejo Au-Ag(-Cu) high-sulfidation deposit, Dominican Republic. The whole-rock chemistry of precursor and variably altered andesite was used to determine the behavior of 64 elements during quartz-alunite, quartz-pyrophyllite, and chlorite-carbonate-kaolinite facies alteration. For high-sulfidation systems, relatively few mass transfer analyses have previously been successfully undertaken due to the difficulty in accurately identifying elements that remain immobile during alteration. This is necessary to remove the effects of significant bulk mass change on whole-rock element concentrations. Elements typically considered to be immobile in other settings where hydrothermal alteration occurs are unsuitable in the high-sulfidation environment, due to the highly acidic fluids responsible for alteration. We show that LREEs are immobile during high-sulfidation-related alteration because of the low stability of LREE complexes in the fluids (lower ionic potential) relative to HREE complexes, and the resultant formation of alteration minerals, such as woodhouseite, which preferentially retain LREEs on a local scale.

The calculated changes in elements during alteration were integrated with changes in mineralogy and mineral chemistry, determined using electron microprobe and laser ablation ICP-MS, to identify the minerals in which elements are hosted in both precursor and altered andesite. Understanding chemical and mineralogical changes that occur during alteration and mineralization provides an insight into the nature of ore-forming fluids and the changes in their physicochemical properties during intense fluid-rock interaction.

Age and Genesis of W–Mo–Cu Mineralization, Gold Hill, Utah

Nathan J. Carey[1], Simon Jowitt[1], Stephanie E. Mills[2]
1. University of Nevada, Las Vegas, Las Vegas, NV, USA, 2. Utah Geological Survey, Salt Lake City, UT, USA

Utah's Gold Hill mining district hosts a Jurassic felsic pluton emplaced into a Mississippian–Pennsylvanian carbonate-dominated sedimentary sequence that is spatially associated with numerous polymetallic (potentially skarn-type) mineral deposits. The timing of mineralization in this district has not previously been constrained, resulting in uncertainty on the relationship between mineralization and pluton emplacement. This study presents new geochronological data for understudied W–Mo–Cu mineralization associated with a Jurassic pluton and outlines key relationships between the geological setting, magma composition, and paleo-fluid evolution of these polymetallic systems.

Six molybdenite samples dated using the Re-Os technique yielded five ages between 156.8 ± 2.2 and 154.4 ± 2.2 Ma that are consistent with existing unpublished zircon U-Pb ages for the proximal pluton (156.1 ± 1.8 Ma). This suggests that the pluton was likely the controlling factor in the generation of the molybdenite and associated mineralization in this area, indicating these are most likely skarn systems. However, an older age of 165.6 ± 2.4 Ma for the sixth low-Re sample provides evidence of multiple pulses of molybdenite mineralization, potentially indicating an earlier non-skarn molybdenite phase of mineralization or that some skarns are associated with a hitherto unidentified earlier phase of intrusion. This study also presents in-depth petrographic analyses from several mines in the district which outline key paragenetic relationships (cross-cutting and overprinting) and mineral assemblages that constrain the timing of alteration and mineralization. These include (1) early actinolite, apatite, diopside, and scheelite(?) skarn alteration, (2) a main molybdenite, chalcopyrite, bornite, pyrite, magnetite, hematite, and calcite phase of mineralization, and (3) post-skarn supergene malachite, azurite, chrysocolla, chalcocite, calcite, and limonite alteration. Future SEM imaging and EPMA will provide further insights into the genesis of W–Mo–Cu mineralization in the Gold Hill district as well as future exploration targeting in similar terranes.

Paragenetic Relationships at the Tuvatu Alkaline Epithermal Gold Deposit, Fiji

Daniel E. Schmidt[1], Katha Pfaff[1], Thomas Monecke[1], Sergio Cattalani[2]
1. Colorado School of Mines, Golden, CO, USA, 2. Lion One LTD, Nadi, Fiji

The Tuvatu alkaline epithermal Au-Te deposit is located on the island of Viti Levu, Fiji. Mineralization occurs in high grade, narrow veins that are hosted by the 4.85 Ma Navilawa monzonite, which forms a subvolcanic intrusion to a partially eroded shoshonitic volcano. The main mineralized zone strikes almost 900 m, has a vertical extent of more than 500 m, and remains open at depth. The deposit is currently under development and contains an indicated resource of 1.0 Mt grading 8.48 g/t Au (274,600 oz Au) and an inferred resource of 1.3 Mt at 9.0 g/t Au (384,000 oz Au) at a cut-off grade of 3.0 g/t Au. Microanalytical work reveals a complex overlap of high- and low temperature processes. Pervasive high-temperature potassic alteration of the Navilawa monzonite is characterized by abundant hydrothermal K-feldspar and secondary biotite. The potassic-altered monzonite is crosscut by first-stage epithermal veins hosting abundant base-metal sulfides. These base-metal rich veins contain euhedral pyrite as well as dendritic aggregates of chalcopyrite, sphalerite, and galena. Mosaic quartz is the principal gangue mineral. Second-stage epithermal veins host the majority of the Au mineralization. Native gold and telluride dendrites occur with roscoelite in bands of mosaic quartz. Relic microtextures suggest that the quartz formed as the result of the recrystallization of non-crystalline, microspherical silica. The microspherical silica is interpreted to have formed under conditions of rapid silica supersaturation caused by the flashing of the hydrothermal fluids, which also triggered gold deposition at far-from equilibrium conditions. Formation of the high-grade vein material was followed by the deposition of chalcedony or euhedral quartz. The third and final stage of epithermal mineralization consists of barren carbonate veinlets. Identification of fluid flashing during deposit formation has important implications for further exploration at Tuvatu and advances the model for how alkaline epithermal deposits form.

Unraveling Bulk-rock Compositional and Metamorphic Mineral Assemblage Zoning at the Laronde Penna Volcanogenic Massive Sulfide Deposit

Miranda R. Lehman[1], Thomas Monecke[1], Nigel Kelly[2], Katharina Pfaff[1], Patrick Mercier-Langevin[3], Annelie Lundström[5], Chris Sherry[4]
1. Colorado School of Mines, Golden, CO, USA, 2. Bruker Nano Analytics, Kennewick, WA, USA, 3. Geological Society of Canada, Quebec, QC, Canada, 4. Terracore Geo, Reno, NV, USA, 5. Minalyze AB, Sävedalen, Sweden

This study introduces a workflow that was developed to identify geochemical and mineralogical gradients in ore deposits by non-destructive drill core analysis. This was achieved by compositional analysis of drill core by continuous XRF core scanning using the device manufactured by Minalyze AB and conducting short- and long-wave hyperspectral imaging of the core using instrumentation at Terracore in Reno, Nevada. Both datasets were co-registered to allow cluster analysis. Based on the clustering, a subsampling strategy was developed to allow further petrographic investigations on representative thin sections and automated mineralogical analysis.

The workflow developed was tested on drill core from the LaRonde Penna deposit in the Archean Abitibi greenstone belt in Quebec. The deposit is a world-class gold-rich volcanogenic massive sulfide deposit which contains an endowment of 71 Mt of ore grading on average 3.9 g/t Au. The ore lenses and their host rocks have been affected by regional greenschist to lower amphibolite facies metamorphism.

The study of core from LaRonde Penna demonstrates that the metamorphoses hydrothermal alteration halo is strongly zoned. Whole-rock geochemical variations along core intersecting the alteration halo and ore zones closely correlate with metamorphic mineral assemblage zoning. The observed variations in the composition of the felsic volcanic rocks in the footwall of the massive sulfide lenses were introduced during hydrothermal alteration as the intensity of fluid-rock interaction was most pronounced in proximity to the ore zones and decreased outwards away from the major zones of hydrothermal upflow. The composition of the hydrothermally altered rocks controlled metamorphic mineral assemblage formation during subsequent metamorphism.

The research has implications to mineral exploration for VMS deposits in high-grade metamorphic terrains as it establishes the link between whole-rock geochemical gradients and metamorphic mineral assemblage variations at greenschist to amphibolite metamorphic grade

Relationship of Magmatic and Ore-forming Processes in Andesitic Volcanoes – Example from the Štiavnica Stratovolcano, Slovakia

Peter Kodĕra[1], Jaroslav Lexa[2], Bertrand Rottier[3], Alexander Kubač[1], Oscar Laurent[4]
1. Comenius University, Faculty of Natural Sciences, Bratislava, Slovakia, 2. Earth Science Institute, Slovak Academy of Sciences, Bratislava, Slovakia, 3. Département de Géologie et Génie Géologique, Université Laval, Québec, QC, Canada, 4. Institute of Geochemistry and Petrology, ETHZ, Zürich, Switzerland

In big, long-living volcanic centres, such as Štiavnica andesitic stratovolcano in the Western Carphatians, there are present diverse styles of ore deposits with a variable metal endowment. Based on LA-ICP-MS, EMPA and microthermometry data from rocks, melt inclusions and fluid inclusions, magmas and most deposits were sourced from an upper crustal (~1 to ~3 kbar) reservoir that was active more than 3 My. Ore deposits were formed during periods of reservoir cooling when the residual melt reached fluid saturation. This melt is represented by evolved rhyolitic melt inclusions of a roughly stable composition, hosted by rock-forming minerals from different stages of magmatic activity. Fluid inclusions in most deposits (except of brines in Cu-Au skarn-porphyries) showed increased B, As, Sb concentrations indicating that the main source of fluids was a magmatic supercritical fluid contracted to liquid during ascent from the reservoir or a contracted vapor resulting from fluid heterogenization. The relatively constant composition of most studied elements, high Cs contents, and agreement with the published fluid-melt fractionation factors indicate a common long-lasting source of magmatic fluids, exsolved from the evolved interstitial melt. The fluids were continuously exsolved and accumulated but liberated during periodical tectonic events. They include migration via contraction fractures (Pb-Zn-Cu stockwork in granodiorite), ring fractures and shear zone induced by a sector collapse of the volcano (early Au-Ag-Pb-Zn-Cu veins), along with porphyry stocks and ring dikes as a precursor of a caldera collapse (Cu-Au skarn-porphyries), and faults of a resurgent horst uplift (late Ag-Au-Pb-Zn-Cu veins). Ore precipitation was triggered by mixing of magmatic and meteoric water (stockwork), boiling of decompressed fluids (early veins), cooling of contracted vapor affected by late fluid heterogenisation (skarn-pophyries) and mixing+boiling (late veins). Our results show that external factors, rather than melt and fluid compositions, primarily control the different metal endowment of the deposits.

Evolution of Magmatic-Hydrothermal Systems Forming Porphyry Copper Deposits: Results of Quartz Solubility Modeling

Thomas Monecke[1], James T. Reynolds[1, 2]
1. Colorado School of Mines, Golden, CO, USA, 2. Fluid Inc., Denver, CO, USA

Quartz solubility calculations in the H_2O-NaCl system were conducted to explain the occurrence of consistent crosscutting relationships between different vein types in porphyry deposits. The model predicts that high-temperature (≥500°C) 'A' vein quartz associated with potassic alteration of the wall rocks precipitates as a result of cooling of the magmatic-hydrothermal fluids. In shallow (≤1.5 km) deposits, rapid decompression may form 'banded' quartz veins. During continued cooling of the magmatic-hydrothermal system, 'B' vein quartz is formed at intermediate temperatures (≥375-500°C). The fluid inclusion inventory of this quartz suggests formation at fluctuating pressure conditions, marking the lithostatic to hydrostatic transition, and the change of wall-rock behavior from ductile to brittle. Textural evidence from many porphyry veins suggests that hypogene sulfide minerals present in 'A' and 'B' veins postdate the quartz, and contacts between quartz and sulfide minerals commonly show dissolution textures. Hypogene sulfide minerals in 'C' veins form at conditions of retrograde quartz solubility, explaining why mineralized 'paint' veins contain little to no quartz. The quartz solubility calculations suggest that 'C' vein formation occurs at temperatures ~375–450°C from low-salinity, single-phase fluids escaping from the lithostatic to the hydrostatic environment. In many deposits, these veins are associated with chlorite, chlorite-K-feldspar, or chlorite-sericite alteration. Late quartz is formed during continued cooling of the hydrothermal fluids at ≤375°C within the single-phase field of the H_2O-NaCl system. This process is responsible for the formation of quartz in 'D' veins and later base-metal-bearing 'E' veins, which are associated with phyllic, advanced argillic, or argillic alteration. An integrated study of the fluid inclusion inventory and cathodoluminescence properties of stockwork vein samples from deposits such as Cöpler in Turkey, Erdenetiin Ovoo in Mongolia, Santa Rita in New Mexico, and Yulong in China confirms the predictions made from quartz solubility modeling.

Assessing the Role of Tectono-magmatic Setting in the Precious Metal (Au, Ag, PGE) and Critical Metal (Te, Se, Bi) Endowment of Porphyry Cu Deposits

Katie McFall[1], Iain McDonald[2], Jamie J. Wilkinson[3]
1. UCL, London, United Kingdom, 2. Cardiff University, Cardiff, United Kingdom, 3. Natural History Museum, London, United Kingdom

Porphyry Cu deposits commonly contain critical and precious metal by-products, including the chalcophile and siderophile elements Au, Pd, Pt, Ag, Te, Se and Bi. These elements partition into residual sulfides during the partial melting of mantle wedge peridotite during subduction, potentially depleting the source magma for subduction-related porphyry deposits. The chalcophile-rich residual sulfides in subduction-modified subcontinental lithosphere are thought to be the source of metals in post-subduction porphyry deposits, and as such these deposits may be more enriched in chalcophile and siderophile elements than subduction-related porphyry deposits. However, many post-subduction deposits have low Au grades and some subduction-related porphyry deposits have high Au and PGE concentrations.

We test the role of tectonic setting in precious and critical metal endowment of porphyry Cu deposits by comparing whole rock assay, PGE data and LA-ICP-MS sulfide trace element data from three case studies with a newly compiled global database. The Skouries Cu-Au-(PGE) porphyry deposit, Greece, and the Muratdere Cu-Au-Mo porphyry deposit, Turkey are both post-subduction; these are contrasted with the El Teniente Cu-Mo porphyry deposit, Chile, which is a classic subduction-related system. We show that post-subduction porphyry deposit sulfides are relatively enriched in Bi, Te and Se compared to sulfides from subduction-related deposits. However, although some critical and precious metals (Ag, Bi and Se) reside in primary sulfide ore minerals, others (Au, Te, Pd and Pt) are hosted in minor accessory minerals. Whole rock data from mineralized samples show that globally both subduction-related and post-subduction deposits can be precious and critical metal enriched, with metal endowment independent of tectonic setting. PGE-enriched porphyry Cu deposits are also enriched in Bi, Te and Au, and semi-metal melts are suggested to play an important role in PGE transport and concentration in porphyry Cu deposits.

Constraints of Magma, Metal and Sulfur Sources Provide Insight into the Formation of the Taca Taca Bajo Cu-Mo-Au Porphyry Deposit (NW Argentina)

Madeleine Ince[1], Steffen Hagemann[1], Marco Fiorentini[1], Tony Kemp[1], Nora Rubinstein[2], Tim Ireland[3], James Banyard[3]
1. Centre For Exploration Targeting, School of Earth Sciences, University of Western Australia, Crawley, WA, Australia, 2. Instituto de Geociencias Básicas, Aplicadas y Ambientales de Buenos Aires (Universidad de Buenos Aires– CONICET),, Buenos Aires, Argentina, 3. First Quantum Minerals, Vancouver, BC, Canada

The Taca Taca Bajo Cu-Mo-Au porphyry deposit is a rare, well-mineralised (2203.3 Mt @ 0.43% Cu, 0.012% Mo, 0.09 g/t Au) back-arc expression located in an early Eocene to Oligocene metallogenic belt and is the largest known Oligocene porphyry deposit in Argentina. Mineralisation at Taca Taca Bajo is spatially and temporally correlated with NE-SW trending rhyodacitic porphyry dikes, which show protracted crystallisation from 30.5 ± 0.3 to 29.0 ± 0.3 Ma (U-Pb SHRIMP; 95% confidence). Hydrothermal alteration can be defined spatially into four zones: potassic (29.1 ± 0.1 Ma; Re-Os molybdenite), early phyllic, late phyllic and supergene (24-21 Ma; Ar-Ar alunite). The Cu sulfides chalcocite, chalcopyrite and locally bornite are predominantly in the early and late phyllic alteration zones within monzogranite (459.4 ± 2.0 Ma; U-Pb SHRIMP; 95% confidence). *In-situ* sulfur isotopes of pyrite reveal mantle-like sulfur isotope $\delta^{34}S$ values (‰), from -1.1 ± 0.3 in the potassic zone to + 1.7 ± 0.1 (2 standard error) in the late phyllic zone, which may indicate a relationship between increasing H_2S/SO_4 ratio and metal endowment. Also, magmatic zircons from the rhyodacitic porphyry dikes display mantle-like $\delta^{18}O$ values of 5.6 ± 0.1 ‰ to 5.9 ± 0.1 ‰, and strongly positive initial εHf= +5.6 ± 0.4 to +7.4 ± 0.2 (2 standard error). These values indicate a dominant contribution from mantle or juvenile crustal source, perhaps with input from older, meta-igneous crust. The copper may be derived from the same source as the rhyodacites, as whole rock geochemistry reveals that unmineralized Oligocene intrusions to the west of the deposit have elevated Cu (ppm). In summary, isotopic and geochemical data from this study suggest that the rhyodacitic porphyry dikes were emplaced from ~31 to 29 Ma, from an underlying Cu-enriched magma chamber with a mixed juvenile and crustal source and mantle-derived sulfur.

Hydrothermal Mineral Systems Related to the Breakup of the Supercontinent Pangea

Mathias Burisch[1, 3], Gregor Markl[2], Jens Gutzmer[3, 1]
1. TU Bergakademie Freiberg, Freiberg, Germany, 2. University of Tuebingen, Tübingen, Germany, 3. Helmholtz-Zentrum Dresden-Rossendorf, Helmholtz Institute Freiberg for Ressource Technology, Freiberg, Germany

Mesozoic hydrothermal systems host the majority of Europe's fluorspar and barite resources as well as significant resources of metals such as Ag, Co, Zn, Pb, Ni and Cu. Their genetic link to extensional tectonics in conjunction with the opening of the North Atlantic has long been suspected, but their spatial and temporal relation to the tectonic evolution of Europe has remained enigmatic.

We compiled 114 published ages from 20 regions related to Mesozoic mineralization across western and central Europe as well as from deposits located in North Africa. Thorough evaluation of available geochronological data for fluorite-barite-Pb-Zn , native-metal-arsenide-carbonate and MVT-type mineralization in Continental Europe and North Africa reveals a distinct, as yet unrecognized, time-space relationship between the distribution of hydrothermal mineral systems and the tectonic evolution of the Tethys-Atlantic-Caribbean rift system. The onset of fluorite-barite-Pb-Zn, native-metal-arsenide and MVT-Pb-Zn mineralization related to the Tethys-Atlantic-Caribbean rift in Continental Europe and North Africa becomes younger with increasing distance to the initial rift axis: from ~230 Ma in proximal, ~195 Ma in intermediate to 140 Ma in distal positions.

We propose that the onset of ore formation in continental rift zones is controlled by mid-crustal fluid release in response to heating and depressurization related to crustal thinning and mantle exhumation. As crustal thinning and mantle exhumation progresses successively from the main rift zones into the continental hinterland, hydrothermal mineralization in distal areas is significantly younger (140-90 Ma) as in proximal areas (230-200 Ma). Based on this assessment we propose the first continental-scale model for the genesis of Mesozoic hydrothermal ore deposits associated to the breakup of the supercontinent Pangea.

Multiple Metal and Fluid Sources Inferred for the Ernest Henry Iron Oxide Copper Gold (IOCG) Deposit, Queensland, Australia: Insights from Magnetite O, Fe, and Ti Isotopes

Christopher Emproto[1], Ryan Mathur[2], Adam C. Simon[1], Ilya N. Bindeman[3], Linda Godfrey[4], Courteney Dnharam[5], Vladimir Lisitsin[5]
1. University of Michigan, Ann Arbor, MI, USA, 2. Juniata College, Huntingdon, PA, USA, 3. University of Oregon, Eugene, OR, USA, 4. Rutgers University, New Brunswick, NJ, USA, 5. Geological Survey of Queensland, City East, QLD, Australia

Iron oxide copper gold (IOCG) deposits are an important source of Cu; however, their formation mechanisms remain contentious. We utilized integrated Fe, Ti, and O isotopic data for magnetite to trace metal and fluid sources for the Ernest Henry IOCG deposit—the largest IOCG-style deposit in the Cloncurry District, Queensland, Australia. Magnetite grains from within and near the Ernest Henry deposit were analyzed for their O, Fe, and Ti isotope compositions, which are reported as δ18O, δ56Fe, and δ49Ti. The δ18O values of samples from the ore zone (+2.98‰; 2σ = 3.05‰) and adjacent unmineralized lithologies (+1.65‰; 2σ = 1.25‰) suggest a magmatic-hydrothermal origin for these magnetites; non-magmatic δ18O values > c. 5‰ may be explained through carbonate assimilation or contamination. However, δ56Fe values for the ore zone (-0.14‰; 2σ = 0.58‰) and off-deposit (-0.01‰; 2σ = 0.44‰) magnetites are generally lower than the typical range (c. +0.06 to +0.49‰) for magmatic(-hydrothermal) magnetite, suggesting non-magmatic Fe input. Ore zone magnetite δ49Ti (+1.72‰; 2σ = 2.23‰) compositions exhibit a loose correlation with δ18O (R^2 = 0.52) and are most variable within the ore zone, suggesting that Ti isotope compositions are not as resilient to post-magmatic processes as has been suggested. The data are best explained by invoking multiple fluids including both magmatic and non-magmatic components. The earliest generation of magnetite formed from components inherited from the host rock protolith(s) wherein magmatic(-hydrothermal) O, Fe, and Ti isotopic compositions are preserved. Later fluid flow channeled between the footwall and hanging wall shear zones introduced non-magmatic Fe that may have been leached from local mafic units by magmatic-hydrothermal fluids, resulting in magnetite with magmatic δ18O, but non-magmatic δ56Fe. Our results demonstrate that multiple fluid and metal sources may contribute to the formation of IOCG deposits.

Exploration Search Space Constraints for Copper-Gold Host Rocks in the Timok Magmatic Complex

Alan J. Wainwright[1], Robert G. Lee[1], Craig J. Hart[1], Richard M. Friedman[2]
1. MDRU-UBC, Vancouver, BC, Canada, 2. UBC, Vancouver, BC, Canada

The Upper Cretaceous Timok Magmatic Complex hosts some of the largest copper-gold resources in the Western Tethyan orogenic collage. However, the timing relationships between mineralization episodes and host volcanic packages are not well constrained. The Timok volcanic, sub-volcanic and volcaniclastic rocks range from basalt to trachyte compositions and are dominantly medium- to high-K calc-alkaline. Mafic- to intermediate volcanic rock suites can be difficult to tell apart in the field but are distinguished using Zr/Ti and Nb/Y plots. Importantly, volcanic facies and composition changes correspond to discrete timing breaks within the stratigraphic sequence. We have characterized key volcanic rock suites from the central and southwest Timok and have constrained their ages using LA-ICP MS (U-Pb zircon).

Data compilation of regional U-Pb zircon grain ages indicate that Timok arc activity began at ~92 Ma with emplacement of an 80 km long by >20 km wide zone of arc magmatism (present coordinates after shortening). Reported major magmatic ages (U-Pb) for the Eastern Timok are ~90 Ma, ~85 Ma and ~82 Ma, and the Re-Os ages are ~88 to ~86 Ma. Our data indicate that the Central Timok volcaniclastic basalt is 78 ± 3 Ma or younger, and the Central Timok volcaniclastic andesite is 79 ± 3 Ma or younger (interpreted maximum ages; LA-ICP MS). We have also detected a mineralized porphyry intrusive in the southwest Timok with an age of 70 ± 1 Ma (LA-ICP MAS; zircon).

Our geochronology and map constraints indicate that 1) volumetrically, most of the Timok stratigraphy at the current erosion level is younger than ~78 Ma and may overlie important host rocks which are only exposed on the Timok margins, and 2) despite the lack of major copper discoveries in <78 Ma host rocks, the Timok continued to yield porphyry-style footprints after ~70 Ma.

Magnetite Chemistry as Exploration Tool to Constrain Fertility and Hydrothermal Evolution in IOCG and Porphyry Environments. San Matías District, Northern Colombia

Julian Manco[1], Craig J. Hart[1], Hildebrando Leal-Mejía[1], Maria A. Rodriguez-Mustafa[2], Farhad Bourzari[1]
1. MDRU-Mineral Deposit Research Unit, The University of British Columbia (UBC), Vancouver, BC, Canada, 2. Department of Earth and Environmental Sciences, University of Michigan, Ann Arbor, MI, USA

Magnetite structurally incorporates minor and trace elements depending on temperature of formation, thus making it useful for petrogenetic interpretations, as an exploration tool, and to understand the physical-chemical conditions of the ore-forming fluid(s). The San Matías Cu-Au District (SMD) in northwestern Colombia hosts multiple late Cretaceous intrusions and associated mineralization, including Fe-Au±Cu±light REE at the El Alacrán IOCG deposit and porphyry Cu-Au deposits like Montiel E and Costa Azul. Much of the mineralization has abundant magnetite, specularite, and mushketovite, particularly at El Alacrán where magnetite is associated with an early calc-silicate alteration. Copper precipitated with later stage sericite-Mg-chlorite-carbonate.

Six hydrothermal magnetite phases (Mag) are recognized from petrography, SEM and EMPA evaluation: Mag 1-intergrown with chlorite, has patchy zonation, elevated Al, Si, and low V; Mag 2-euhedral, dark-gray and apatite inclusions, highest Al, Mg, Si and Ca; Mag 3-unzoned, euhedral, light-gray, lacks apatite inclusions, highest Fe and lowest Al, Mg, and Ca; Mag 4-pyrite-Au-related, bladed (mushketovite), high Al, Mg, Si, Ca and low V; Mag 5-mushketovite with inherited specularite, lowest Ti and Ca; Mag 6-unzoned and within quartz veinlets in sericite-Mg-chlorite-carbonate alteration.

Mag 6 from the El Alacrán deposit contains Ca-Al-Mn-Ti-V concentrations similar to magnetite from quartz-rich tonalites intrusions and to the hydrothermal magnetite from the associated porphyry deposits. El Alacrán magnetites record an initial heating path from Mag 1 to Mag 2 with a subsequent cooling path to Mag 3, 4 and 5. The changes in the chemistry of magnetite reveal the thermal history related to early calc-silicate alteration (Mag 1 – 5). The similar composition of Mag 6 with the magnetite from the porphyry prospects enhances the prospectivity for a potential concealed porphyry under El Alacrán.

Assessing Proximity to Iron Oxide-Copper-Gold (IOCG) Mineralization Using Monazite Chemistry

Caroline Tiddy[1], David Giles[1], Adrienne Brotodewo[1], Diana Zivak[2], June Hill[3], Jim Hodgkison[4], Mitchell Neumann[4]
1. MinEx CRC, Future Industries Institute, University of South Australia, Adelaide, SA, Australia, 2. Department of Earth Sciences, University of Adelaide, Adelaide, SA, Australia, 3. CSIRO Mineral Resources, Perth, WA, Australia, 4. OZ Minerals Limited, Adelaide, SA, Australia

Geochemical indicators of proximity to mineralisation have been proposed for a variety of minerals (e.g., chlorite/epidote/micas for porphyry systems: Cooke et al. 2021: Econ. Geol.). Monazite is a mineral that has not received much attention in this space but is interesting to consider as it is a resistate mineral phase, meaning it can withstand processes of weathering and transport. Monazite has been demonstrated to preserve chemical compositions that can be linked to iron oxide-copper-gold (IOCG) mineralization at Prominent Hill in the northern Gawler Craton, South Australia, and that is distinguishable from the chemistry of monazite in surrounding unmineralized basement rocks that grew in association with other geological processes (e.g., metamorphism) (Forbes et al. 2015: J. Geochem. Expl.). The limitation of Forbes et al. (2015) is that the study was restricted to one deposit.

In this study, we assess the chemistry of hydrothermal monazite from the Carrapateena IOCG deposit in the central Gawler Craton. We compare the Carrapateena monazite chemistry to metamorphic monazite and monazite grown in association with shear zones. Carrapateena monazite is characterised by elevated light rare earth element (LREE) and depleted Y and Th concentrations, which is comparable to observations from the Prominent Hill deposit. Additionally, the Carrapateena monazite is depleted in Nd. The observation of similar monazite chemistry in multiple IOCG deposits is used here to propose broad-scale criteria for IOCG exploration within the Gawler Craton.

Monazite hosted within younger cover sediments overlying the mineralized basement rocks at both Prominent Hill and Carrapateena also preserves the chemical signature of IOCG-related monazite. These observations suggest that monazite chemistry is preserved during weathering, erosion, transport and redeposition processes. The IOCG-related chemical signature of monazite therefore has the potential to be dispersed within cover sequence materials, effectively increasing the geochemical footprint of mineralization.

High-REE Mafic Magmatism in the Mojave: The Shonkinite Connection Between Mountain Pass and Bobcat Hills, California

Kathryn E. Watts[1], David M. Miller[2]
1. U.S. Geological Survey, Spokane, WA, USA, 2. U.S. Geological Survey, Moffett Field, CA, USA

The Bobcat Hills are located ~55 km southeast of the Mountain Pass rare earth element (REE) mine in the Mojave Desert of southeastern California. Recent mapping at Bobcat Hills has delineated a shonkinite dike swarm that correlates to the Mountain Pass system. The dikes crosscut Paleoproterozoic gneiss host rocks. Over a dozen dikes, most 1-2 m wide, are similar to Mountain Pass shonkinites, with a dark gray groundmass and conspicuous black phlogopite crystals. Whole-rock geochemistry demonstrates a chemical affinity to Mountain Pass shonkinites, with 50-52 wt % SiO_2, 8-10 wt % K_2O, 10-12 wt % MgO, 1-2 wt % F, and light REEs up to ~1,000x chondritic. Chemical differences include lower Ba (<0.5 wt %) and Sr (<700 ppm) for Bobcat Hills shonkinites. Zircons from four shonkinite dikes yielded error-weighted mean $^{207}Pb/^{206}Pb$ ages and paired concordia $^{206}Pb/^{238}U$-$^{207}Pb/^{235}U$ ages of 1425-1428 ± 2-3 Ma. These ages are on the older end of the zircon age range determined for Mountain Pass shonkinites (~1430-1400 Ma). The Bobcat Hills shonkinites lack inherited Paleoproterozoic zircons indicative of crustal assimilation, which are ubiquitous in Mountain Pass shonkinites. This may be due to relatively small-volume and ephemeral shonkinite intrusions in cold crust (Bobcat Hills) vs. more voluminous and protracted shonkinite intrusions in warm crust primed for assimilation (Mountain Pass). Zircon trace elements are similar, with low Hf (<10,000 ppm), high Eu/Eu* (~0.8-1.3), and high Th/U (>1.5). Bobcat Hills zircons have the highest Th (>2,000 ppm) and Th/U (3-5). Zircon Ti concentrations (~50-200 ppm) and crystallization temperatures (~900-1,100 °C) overlap to exceed Ti concentrations (<100 ppm) and crystallization temperatures (<1,000 °C) for Mountain Pass zircons. Bobcat Hills may be a critical locality for understanding mafic Mesoproterozoic magmatism in the region and mantle sources of REEs that led to the formation of the world-class Mountain Pass REE deposit.

Rare Earth Elements in Coal Combustion Byproducts of Saskatchewan: a Geochemical, X-ray Adsorption Spectroscopy, and Machine Learning Investigation

Brendan A. Bishop[1], Karthik Ramachandran Shivakumar[2], Jamie Schmidt[1], Daniel S. Alessi[2], Leslie J. Robbins[1]
1. University of Regina, Regina, SK, Canada, 2. University of Alberta, Edmonton, AB, Canada

Rare earth elements (REE) are essential for the global transition to clean energy sources owing to their role in high powered magnets used in technologies such as wind turbines and electric vehicles. Therefore, demand for REE is expected to increase in the coming years, however current North American production is not sufficient to meet this anticipated demand. This lack of domestic production could have significant economic and national security implications and negatively impact our ability to meet climate change goals. Recently, there has been increased interest in extracting REE from secondary sources, such as coal combustion byproducts (CCBs) including fly ash and bottom ash, however their potential as a REE source is dependent on the geochemistry of the particles which must be understood before extraction strategies can be developed.

This contribution investigates REE in CCBs from three power plants which burn lignite coal in southern Saskatchewan to ascertain the variables which affect the potential for REE recovery. Total digestions were performed to determine REE abundances, sequential extractions were used to elucidate the dominant REE host phase, and acid leaches were utilized to quantify the leaching potential of the CCBs. These data were then combined with existing geochemical characterizations of CCBs from across North America and subjected to machine learning algorithms and statistical analyses to determine elemental associations and geochemical indicators of REE enrichment. Finally, the molecular binding environment of REE was determined using X-ray adsorption spectroscopy (XAS). This study is the first step in assessing the potential for REE extraction from CCBs from Saskatchewan, while insights from this study can be applied to CCBs in other localities. Identifying new, environmentally friendly sources of REE is crucial in order to meet climate goals and to secure energy independence.

An Abrupt Switch in Magmatic Plumbing Taps Porphyry Copper Deposit-forming Magmas

Lawrence C. Carter[1], Simon R. Tapster[2], Ben J. Williamson[1], Yannick Buret[3], David Selby[4], Gavyn K. Rollinson[1], Ian Millar[2], Daniel B. Parvaz[5, 6]
1. Camborne School Of Mines, Corwall, United Kingdom, 2. British Geological Survey, Nottingham, United Kingdom, 3. Natural History Museum, London, United Kingdom, 4. Durham University, Durham, United Kingdom, 5. Selfrag AG, Kerzers, Switzerland, 6. Lightning Machines, Manchester, United Kingdom

Porphyry-type deposits are a vital source of green technology metals such as copper and molybdenum. They typically form in subduction-related settings from large, long-lived magmatic systems. The most widely accepted model for their formation requires that mantle-derived magmas undergo a ramp-up of volatiles and ore-forming constituents in mid- to lower crustal reservoirs over millions of years. However, these models are based on partial records of the shallow level rock record of porphyry deposits. To address this, we have evaluated the timeframe and geochemical signatures of magmatic rocks in a ~8 km palaeodepth cross-section through the classic Yerington magmatic system, Nevada, which is particularly well exposed in 3D and has provided constraints for most modern porphyry exploration models. Here we show that the magmas underwent a major and abrupt change in chemistry over a period of <200 kyrs that is coincident with the initiation of ore formation. This change in magma chemistry, which is documented across the plutonic to volcanic record, is attributed to a switch in the nature of the magmatic plumbing system. It is suggested that volatile-rich granitic melts were extracted from an estimated ~30 km depth and transported to shallow levels (~3-8 km) where they evolved to exsolve porphyry mineralising fluids. These were focused upwards through highly permeable crystal-mush dyke pathways. The rapid timescale of the switch suggests that the increase in a magma's ore-forming potential is not solely driven by a progressive ramp-up in ore-forming constituents, due to tectonic controls over multi-million year time scales, but through processes within the melt evolution zone, operating more than an order of magnitude faster than previously envisaged. This short timescale, in which geochemical signatures associated with mineralisation emerged, narrows the temporal-geochemical footprint of magmas associated with porphyry mineralisation, that can aid in exploring for new porphyry-type ore deposits.

Twin Canyon, Gold Mineralization in a Hydrocarbon Reservoir

Jon P. Thorson[1], Poul Emsbo[2], Jay M. Thompson[2], Heather A. Lowers[2]

1. Consulting Geologist, Denver, CO, USA, 2. US Geological Survey, Denver, CO, USA

Twin Canyon gold prospect, Montezuma County, Colorado, is located on the southwest flank of the La Plata Mountains in Jurassic eolian Junction Creek Sandstone. The prospect is located in the crest of an anticlinal structure in fine-grained bleached and altered sandstone sealed above by the Morrison Formation. Detrital feldspar in the sandstone was altered to kaolinite, and calcite cement has been dissolved leaving the host rock soft and friable. Gold is intimately associated with bituminous hydrocarbons. Bitumen occurs as disseminated spots, dense plumes of concentrations of spots, concretionary masses, and vein and fracture fillings. Bitumen is associated with native gold and has an uncertain paragenetic relationship with disseminated iron oxides, which occur in deformation bands, concretionary masses, and as selvedges around plumes of bitumen concentrations.

Two sets of veins and fractures occur in the Charlene Mine. Early swarms of discontinuous deformations bands accentuated with concentrations of gray to brown limonite and containing gray calcite centers are crosscut by veins and fractures filled with friable dark-colored bitumen with occasional central open spaces lined with fine calcite crystals. Bitumen veins are also characterized by dense halos of disseminated bitumen that extend into the wall rock sandstone.

Historical channel sampling in the underground prospect has returned assays up to 8 gm Au/ton over 3 m. Historical drilling indicates multiple stratiform mineralized horizons within the Junction Creek Sandstone. Sampling along the projected Junction Creek Sandstone has anomalous gold in soils (> 20 ppb Au, up to 500 ppb Au) for over 3 km across the axis of the anticline. Native gold occurs as fine flakes and grains disseminated in sandstone and in bitumen-filled veins and fractures. Laser ablation inductively coupled mass spectrometry and electron microprobe analyses indicate the bitumen contains as much as 50 ppm "no-see-um" gold.

Fluid Evolution and Ore Genesis of the Permian Hongshanliang Manto-type Copper Deposit in the Eastern Tianshan

Liandang Zhao[1], Huayong Chen[2], Jinsheng Han[3]
1. School of Earth Science and Resources, Chang'an University, Xi'an, China, 2. Key Laboratory of Mineralogy and Metallogeny, Guangzhou Institute of Geochemistry, Chinese Academy of Sciences, Guangzhou, China, 3. School of Earth Resources, China University of Geosciences, Wuhan, China

The Hongshanliang copper deposit, a typical copper deposit in the Aqishan-Yamansu belt, Eastern Tianshan, is hosted in the tuff of the Lower Carboniferous Yamansu Formation. Five alteration/mineralization stages have been established: chlorite–sulfde stage (Stage I), quartz–pyrite stage (Stage II), quartz–polymetallic sulfde stage (Stage III), late veins (Stage IV), and supergene process (Stage V), based on detailed crosscutting relationships of veins, textural relationships, and mineral assemblages. Stage I has chlorite–pyrite ± chalcopyrite ± quartz ± sericite assemblage and is commonly characterized by euhedral pyrite with silicates and/or chalcopyrite as its pressure shadow, indicating an early mineralization event with apparent deformation. The late main mineralization stage (Stage III) has typical mineral assemblages of quartz–chalcopyrite ± chlorite ± calcite ± sericite, quartz–chalcopyrite–sphalerite–galena ± pyrite, and quartz–sphalerite–galena ± chlorite, with veins, veinlets, disseminations, and local massive ore types, which is similar to Manto-type copper deposits regarding mineralization and ore structures.

Detailed fluid inclusion study shows an evolved hydrothermal ore-forming system, with temperatures decreased from Stage II (310–480 °C) through Stage III (160–180 °C) to Stage IV (130–170 °C), and corresponding salinities of 2.7–26.2 wt% NaCl equiv., 2.1–12.3 wt% NaCl equiv., and 0.5–7.7 wt% NaCl equiv., respectively. Such fluid evolution is consistent with corresponding stages of H–O isotopes. Moreover, sulfur isotopes suggest that the two mineralizing fluids are predominantly magmatic–hydrothermal, with other influx contribution into the ore-forming system, i.e., minor Early Carboniferous seawater in Stage I and minor organic materials in Stage III.

In combination of alteration, paragenesis, nature and source of ore-forming fluids, sericite $^{40}Ar/^{39}Ar$ geochronology, and comparison with other typical deposits, we proposed that the Hongshanliang copper deposit underwent two mineralization events with the Permian main mineralization similar to Manto-type copper deposits.

Evaluation of the Tungsten Potential in the Eastern Alps – New Insights from Geological, Mineralogical and Chemical Data on Scheelite

Florian Altenberger[1], Johann G. Raith[1], Joachim Krause[2], Kai Bachmann[2], Julia Weilbold[3], Holger Paulick[3]
1. Montanuniversität Leoben, Department Applied Geosciences and Geophysics, Chair of Resource Mineralogy, Leoben, Austria, 2. Helmholtz-Zentrum Dresden-Rossendorf, Helmholtz Institute Freiberg for Resource Technology, Freiberg, Germany, 3. Geological Survey of Austria, Vienna, Austria

Tungsten deposits and showings were investigated from different geological units of the Eastern Alps using a combination of cathodoluminescence (CL), electron probe micro-analyzer, SEM-based automated mineralogy and laser ablation-inductively coupled plasma-mass spectrometry. In the Eastern Alps, scheelite ($CaWO_4$) is the main carrier of tungsten, one of the prominent raw materials highlighted by EU criticality assessments for future technologies. The W-Alps project aims to develop assessment criteria for the evaluation of regional tungsten potentials in Austria and includes field-based and chemical investigations of several tungsten occurrences reflecting different types of mineralization to constrain their genesis. Among these, the world-class Felbertal W-deposit is the economically most important. The petrography and mineral chemistry of its scheelite is compared to scheelite from other occurrences to indicate the tungsten endowment. Our data demonstrate that the quest for "primary" scheelite signatures is complicated by coupled dissolution-reprecipitation and recrystallization processes during hydrothermal formation but also subsequent regional metamorphism, causing multiple types of scheelite with a broad chemical composition in most occurrences. It becomes clear that a few random scheelite analyses are insufficient to unravel the origin of an ore deposit and can lead to wrong interpretations. Thus, scheelite types in all occurrences were chronologically sorted by combining geological information, CL-textures and distinct trace element patterns. The distribution of trace elements in scheelite is mainly controlled by the fluid chemistry, element fractionation during crystallization of scheelite and co-existing phases and is also significantly modified during subsequent metamorphic mobilization of pre-existing scheelite. We will point out this temporal evolution of scheelite textures and chemistry and its importance for genetic implications and mineral exploration.

Tracking Sediment Provenance in a River Basin for Mineral Exploration: the Sediment Fingerprinting Technique

Jorge Crespo[1], Elizabeth Holley[1], Katharina Pfaff[2], Madeleine Guillen[3], Ivan Lizaga[4], Sergio Ticona[3], Isaac Simon[2]
1. Colorado School of Mines, Department of Mining Engineering, Golden, CO, USA, 2. Colorado School of Mines, Department of Geology and Geological Engineering, Golden, CO, USA, 3. Universidad Nacional de San Agustin de Arequipa, Departamento de Geologia y Geofisica, Arequipa, Peru, 4. Ghent University, Department of Green Chemistry and Technology, Isotope Bioscience Laboratory - ISOFYS, Gent, Belgium

Stream sediment geochemical surveys are a standard exploration tool, but the results can constitute mixing problems that are challenging to solve. Sediment data record water-rock interaction in tributaries followed by fluid mixing in larger downstream catchments, potentially obfuscating the relative trace element contributions of each tributary. We propose a workflow using FingerPro, a state-of-the-art unmixing model implemented as an R package, to quantify the sediment source contribution in order to vector towards the source of anomalous values during exploration.

We present an example workflow from the Ocoña basin, one of the most important drainages in the southern part of Peru. The basin hosts more than 70 recognized Au-Ag and base metal deposits, and exploration is increasing in the region. We applied an unmixing model algorithm to a geochemical database of stream sediment samples from IMGEMMET-Peru. The FingerPro package unmixes the samples after selecting a set of optimal tracers that best fit the reality of our database. This study selects three major subcatchments as potential sediment sources (Upper, Middle and Lower). As a target, we use two mixture samples located in the lower part of the Ocoña basin. The tracers that best fit the model were: Co, As, Cu, Pb, and V based on the consensus and consistency methods. Our preliminary results suggest that the main sediment contributing source is the upper part of the basin, followed by the middle part of the basin, with the least contribution from the lower part of the basin.

Based on these results, we can rule out the lower part of the Ocoña basin as a possible source of noble and base metals. The areas with the greatest potential for mining exploration correspond to the middle and upper part of the catchment, where most small-scale mining and exploration work is currently being carried out.

Pyrite Geochemistry as a Vector Toward Mineralization in Irish-type Zn-Pb Deposits

Claire Geel[1], Koen Torremans[1], Aileen Doran[1], Oakley Turner[1], Tadgh Dornan[2], Murray Hitzman[1]
1. iCRAG SFI Research Centre in Applied Geosciences, Dublin, Ireland, 2. Trinity College Dublin, Dublin, Ireland

Pyrite is the most abundant sulfide mineral in the Earth's crust and is commonly associated with economic mineralization. The geochemistry of pyrite can record the conditions of orebody formation and the composition of associated fluids and may provide a vector toward mineralization. Irish-type Zn-Pb mineralization was facilitated through the mixing of hot metal-bearing hydrothermal fluids and cold hypersaline brines with bacteriogenically reduced sulfur. Associated pyrite occurs in both non-economic and economic zones and their trace element ratios such as Co/Ni; As/Cu; Tl/Fe can act as fingerprints of hydrothermal fluids, thereby aiding in estimating the quality and tenor of nearby ore. Therefore, pyrite trace elements may form geochemical halos around Zn-Pb orebodies similar to other metal distribution haloes in Ireland. Additionally, hydrothermal pyrite grains also display zonation with geochemical variation such as increasing Ag, As and Tl that records changes in the fluid composition from core to rim related to hydrothermal pulses. From previous work, we know those trace elements incorporated into the pyrite crystal lattice record different element assemblages depending on whether they are hydrothermally sourced (As-Cu-Co-Tl-Zn enriched) or derived from bacterial sulfate reduction (Mo-Mn-Ni-Se enriched), as determined from in-situ $\delta^{34}S$ measurements. In this study, multidimensional multimethod statistical classification of pyrite is applied to new and historic data. Laser ablation-inductively coupled plasma-mass spectrometry (LA-ICP-MS) and whole-rock lithogeochemistry data sets, collected from drillcores positioned both proximal and distal to orebodies in the Irish Midlands, are analysed using Multivariant Analysis of Variance, Correlation Cluster Analysis, Principal Component Analysis and Factor Analysis. Additionally, both Logistic Regression and Random Forest machine learning algorithms will be applied, building on methodologies developed for fingerprinting pyrite in building aggregates in Ireland. These results will further inform the generation of a systematic standardized method for measuring trace element geochemistry from pyrite in different orebodies and mineral exploration cores.

Remote Sensing of Gudjareti- Khachkovi Gold Ore Field. Adjara-Trialeti Folded Zone, Lesser Caucasus (Georgia)

Giorgi Mindiashvili
Ivane Javakhishvili Tbilisi state university, Tbilisi, Georgia

This research consists of interpretations of more than 40 data obtained by remote sensing analyses and techniques in order to detect geological features and potential of probable Gudjareti- Khachkovi ore fields in the area located in South-East Georgia (Adjara-Trialeti folded zone, Lesser Caucasus) using Terra ASTER Multispectral satellite data.

Technical specifications for the ASTER data used are as follows:

Granule ID: AST3A1 0409160806091107270062

Processing Level: 3

Acquisition Date: 20040916

Scene ID: [171, 87, 1]

Processed Bands: "01023N3B0405060708091011121314"

Cloud Coverage: 2

The area is quite intensive in terms of vegetation, snow, glacier, and cloudiness. Therefore, the best satellite data among all ASTER images have been ordered that could represent the area the best way and which has the least snow and cloud cover. The acquisition date of the satellite data is 24.08.2012. It is noticeable that although the time of acquisition is autumn season, excess snow covers still remain on mountain tops. The ASTER image used in this study has approximately 2% cloudiness, 20% vegetation, and 35% snow cover in the whole region. In this very highly vegetated territory, there have been detected different rock types covering 10% of the study area and lithological descriptions, mineral mapping techniques, and analysis towards determination of geological features have been made.

AUTHOR INDEX